AF896157

Biomass Densification

Jaya Shankar Tumuluru

Biomass Densification

Systems, Particle Binding, Process Conditions, Quality Attributes, Conversion Performance, and International Standards

Springer

Jaya Shankar Tumuluru
Biological Processing Department
Idaho National Laboratory
Idaho Falls, ID, USA

ISBN 978-3-030-62887-1 ISBN 978-3-030-62888-8 (eBook)
https://doi.org/10.1007/978-3-030-62888-8

© Springer Nature Switzerland AG 2020
This work is subject to copyright. All rights are reserved by the Publisher, whether the whole or part of the material is concerned, specifically the rights of translation, reprinting, reuse of illustrations, recitation, broadcasting, reproduction on microfilms or in any other physical way, and transmission or information storage and retrieval, electronic adaptation, computer software, or by similar or dissimilar methodology now known or hereafter developed.
The use of general descriptive names, registered names, trademarks, service marks, etc. in this publication does not imply, even in the absence of a specific statement, that such names are exempt from the relevant protective laws and regulations and therefore free for general use.
The publisher, the authors, and the editors are safe to assume that the advice and information in this book are believed to be true and accurate at the date of publication. Neither the publisher nor the authors or the editors give a warranty, expressed or implied, with respect to the material contained herein or for any errors or omissions that may have been made. The publisher remains neutral with regard to jurisdictional claims in published maps and institutional affiliations.

This Springer imprint is published by the registered company Springer Nature Switzerland AG
The registered company address is: Gewerbestrasse 11, 6330 Cham, Switzerland

I dedicate this book to my wife, Mrs. Naga Sri Valli Gali, and two daughters, Priya Lasya Tumuluru and Siva Lalasa Tumuluru.

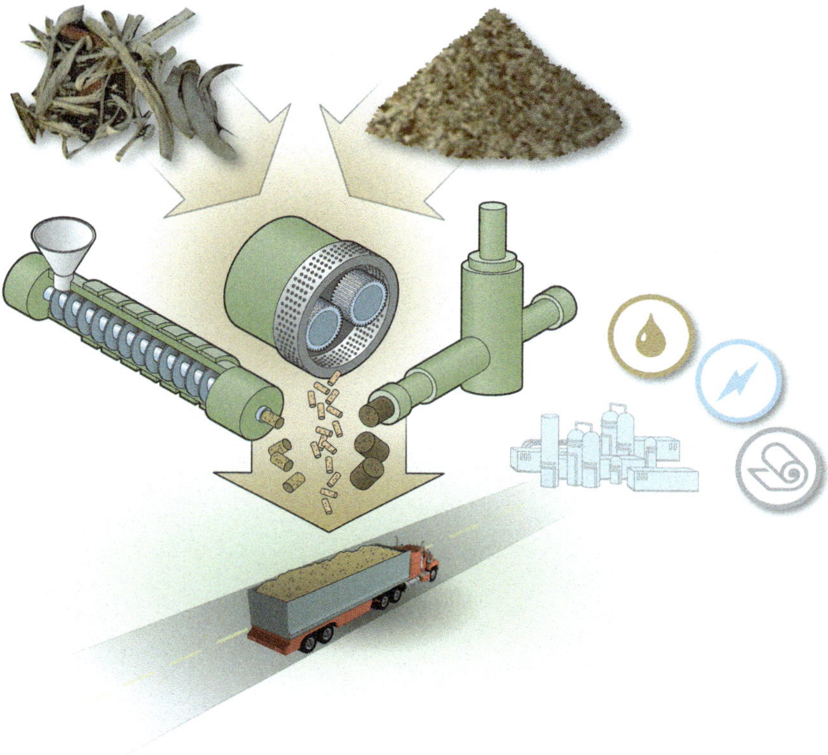

Preface

Biorefineries primarily use woody, herbaceous, and municipal solid waste (MSW) biomass for biofuels and biopower production. Biomass is considered a renewable energy source as it is carbon neutral, can be regrown in short amounts of time, and the CO_2 released during its conversion to energy is naturally a part of the carbon cycle. Of the total amount of biomass fuel used to create primary energy in the United States in 2016, 48% came from ethanol, 41% from woody and wood-derived biomass, and the remaining 11% from MSW. According to the US Billion-Ton Report, there are more than one billion tons of biomass available for bioenergy applications in the United States. Among the many types of biomasses used for biofuels production—forest-based biomass; agricultural residues, such as corn stover; and dedicated energy crops, such as switchgrass and miscanthus—all have low-density and variable moisture content, which create handling and storage issues and limit the viability of long-distance transportation. The density of wood chips and ground herbaceous and MSW biomass is typically about 5–15 pounds per cubic foot, depending on the grind size, feedstock type, and moisture content. Among the various biomasses available, woody biomass chips have a higher density, as well as a definite size and shape. Increasing the density of agricultural and corn residues to >25–45 lb/ft^3 is critical in handling large volumes. The density requirement of biomass changes based on the mode of transportation used (i.e., transporting by truck is weight-limited, whereas transporting by rail or ship is volume-limited). For weight-limited systems, such as trucks, a maximum density of 20–25 lb/ft^3 can fill the truck completely full. Selecting the right type of densification system—as well as process variables to meet the density requirements for different modes of transportation—will be critical to reducing densification costs and increasing transportation efficiencies.

Biorefineries are currently unable to use all the billion tons of biomass available in the United States for bioenergy applications due to issues such as irregular size and shape, low density, lower energy content, and the chemical and biochemical composition of the biomass. Currently, the biggest challenges that biorefineries face are in transporting biomass efficiently, handling it, and finding a way to feed it through the system in a more uniform manner. Indeed, biorefineries around the world have incurred major challenges in terms of their inability to handle ground biomass due to its moisture and particle size variability and the chemical and biochemical changes the biomass undergoes during storage issues, as well as the great challenges inherent in transporting low-density biomass over longer distances. The good news is that many of these challenges can be overcome by mechanical and thermal preprocessing and pretreatment technologies. Mechanical preprocessing technologies such as grinding and densification can help produce an aerobically stable high-density product. One of the major challenges with this strategy is that biomass densification technology is currently cost- and energy-intensive. The future success of the biofuels industry will depend on the development of other ways to densify biomass economically. Densification converts biomass material into something much more suitable as a commodity product. The largest advantages of densification are: (a) improved handling and conveyance efficiencies throughout the supply system and biorefinery infeed; (b) controlled particle size distribution for improved feedstock uniformity and density; (c) fractionated structural components for improved compositional quality; and (d) conformance to predetermined conversion technology and supply system specifications. Densification systems commonly used for biomass include: (i) a pellet mill; (ii) a cuber; (iii) a briquette press; (iv) a screw extruder; (v) a tablet press; and (vi) an agglomerator. These systems are adapted from the food, feed, and pharma industries. Among these, the pellet mill, briquette press, and screw extruder are the most common systems used for bioenergy production. The quality of densified biomass produced using these systems is evaluated with the existing international standards developed for pellet mill and briquette press systems; there are no systems-specific standards developed for the others.

Different densification systems are widely available. This book will focus on: (a) how these densification systems impact the quality of the densified products; (b) how each of the densification system variables impacts cost and quality; (c) how the different biomass properties impact quality; and (d) the different densification process models, as well as the process optimization and international standards for densified biomass. Another emphasis of the book is the advancement of densification technologies and the process used to reduce the costs of the densified biomass to make densification a variable technology regarding cost, quality, and conversion performance for the biofuels industry.

Organization of the Book

Based on the subject matter, the book has six chapters. The first chapter focuses on different feedstocks and their physical properties, chemical and biochemical composition, and their importance for biofuels production. The second chapter focuses on densification systems, binding mechanisms, process variables, and quality attributes, with an emphasis on how the quality of the densified products is influenced by the process variables and densification systems. The third chapter focuses on the densification process models, the scale-up methodology, and the optimization of the process using statistical and novel evolutionary algorithms. The fourth chapter discusses the advancements in the biomass densification process and the positive impact these advancements generate on the overall cost and quality. The novel preprocessing technologies that are developed at Idaho National Laboratory (INL), such as fractional milling, high-moisture pelleting, and low-temperature drying, are also discussed in this chapter, as well as a discussion about efficient moisture management in the biomass using dewatering technologies and advanced comminution technologies to reduce moisture removal costs from the biomass. The fifth chapter discusses the European, international, and American standards that are available for evaluating the quality of the densified products, as well as a comparison of the quality of pellets and briquettes produced using a pellet and a briquette press at different stages of the process. Finally, the last chapter discusses the suitability of the resulting pellets and briquettes for biochemical, thermochemical, and biopower applications.

Who Will Benefit from This Book?

This book will help biomass engineers, biorefineries managers, and academic researchers to understand how mechanical methods, such as size reduction dewatering and densification, can help to overcome some of the biomass limitations, in terms of physical properties such as low density, variable moisture, irregular size and shape, and morphological characteristics, as well as a discussion on how the densified products impact the biochemical and thermochemical conversion. This book will also help science and engineering students understand why biomass needs to undergo mechanical preprocessing for producing a uniform, stable, and densified product for biofuels applications. Further, the book will help policymakers and energy systems planners in creating awareness on how biomass limitations in terms of density, size, shape, and moisture should be overcome to make biorefineries successful.

US Department of Energy Disclaimer

This information was prepared as an account of work sponsored by an agency of the U.S. Government. Neither the U.S. Government nor any agency thereof, nor any of their employees, makes any warranty, express or implied, or assumes any legal liability or responsibility for the accuracy, completeness, or usefulness of any information, apparatus, product, or process disclosed, or represents that its use would not infringe privately owned rights. References herein to any specific commercial product, process, or service by trade name, trademark, manufacturer, or otherwise do not necessarily constitute or imply its endorsement, recommendation, or favoring by the U.S. Government or any agency thereof. The views and opinions of authors and editor expressed herein do not necessarily state or reflect those of the U.S. Government or any agency thereof.

Idaho Falls, ID, USA Jaya Shankar Tumuluru

Acknowledgments

The author would like to thank Dr. Richard Hess, INL Bioenergy Program Relationship Manager, Bioenergy Technologies Office, U.S. Department of Energy; Dr. Fredrick Stewart, INL Biological & Chemical Science & Engineering Department Manager; and Dr. Vicki Thompson, INL Group Lead, Biological Processing, for their support in getting this book published. The author would also like to sincerely thank Mr. David Combs, INL Art Director and Branding Specialist, and Mr. Gordon Holt, INL Senior Editor, for their invaluable support in publishing this book. Finally, the author would like to express extreme gratitude and thanks to his family for their help and support.

This work was supported by the U.S. Department of Energy, Office of Energy Efficiency and Renewable Energy under DOE Idaho Operations Office Contract DE-AC07- 05ID14517. Accordingly, the U.S. Government retains and the publisher, by accepting the article for publication, acknowledges that the U.S. Government retains a nonexclusive, paid-up, irrevocable, worldwide license to publish or reproduce the published form of this manuscript or allow others to do so, for U.S. Government purposes.

Contents

1 Bioenergy Feedstock Types and Properties 1
1.1 Introduction 1
1.1.1 Biomass 1
1.1.2 Various Biomass Types According to the International Standards Organization (ISO-17225-1: 2014) 2
1.1.3 Fruit Biomass 3
1.1.4 Converting Biomass 4
1.2 Biomass Properties 5
1.2.1 Proximate and Ultimate Composition 5
1.2.2 Ash Content 8
1.2.3 Calorific Value 9
1.2.4 Biochemical Composition 11
1.2.5 Moisture 13
1.2.6 Bulk Density 13
1.2.7 Biomass Limitations 15
1.3 Conclusions 18
References 18

2 Binding Mechanism, Densification Systems, Process Variables, and Quality Attributes 23
2.1 Introduction 23
2.1.1 Densification 24
2.1.2 Bonding Mechanism During Densification 24
2.1.3 Densification Systems 29
2.1.4 Feedstock Properties 49
2.1.5 Biomass Composition 50
2.2 Conclusions 57
References 58

3 Densification Process Models and Optimization 63
3.1 Introduction 63
3.1.1 Compression Models 64
3.1.2 Application of the Compression Models for Process Scale-Up 71
3.1.3 Testing the Scale-Up Model in a Pilot-Scale Ring Die Pellet Mill 76
3.1.4 Response Surface Models and Optimization of the Pelleting Process 77
3.2 Conclusions 81
References 82

4 Advances in Biomass Densification and Its Impact on Cost and Quality 85
4.1 Introduction 85
4.1.1 Techniques to Overcome the Particle Handling Problems 86
4.2 Conclusions 119
References 120

5 Biochemical and Thermochemical Conversion Performance of Densified Products for Biofuels Production 123
5.1 Introduction 123
5.1.1 Biochemical Conversion 125
5.1.2 Densification Impact on the Sugar Yields 125
5.1.3 Thermochemical Conversion 136
5.2 Conclusions 157
References 157

6 International Standards for Densified Products 161
6.1 Introduction 161
6.2 Global Wood Pellet Consumption 163
6.3 Pellet Quality and Standards 164
6.4 International Standards 166
6.5 CEN Standards 166
6.6 European Common Standard 168
6.7 ISO Standards 172
6.7.1 Comparison of CEN, ISO, and PFI Standards 174
6.8 Conclusions 182
References 187

Index 189

Abbreviations

%	percent
%, w.b.	percent wet basis
V_R	volume ratio
ρ_f	particle rearrangement
°C	degrees Celsius
μm	micron
ACO	ant colony algorithm
AFEX	ammonia fiber expansion
BTU	British thermal unit per pound
C	carbon
Ca	calcium
CBP	conventional briquetting process
CBU	cellobiase unit
CCE	carbon conversion effciency
CDP	continuous dewatering press
CEN	European Committee for Standardization
CGE	cold gas efficiency
Cl	chlorine
CO_2	carbon dioxide
CPP	conventional pelleting process
CrI	cellulose crystallinity index
CT	computed tomography
CV	calorific value
DA	dilute acid
DDGS	dried distillers grains with solubles
DMA	dynamic mechanical analysis
DOE	U.S. Department of Energy
DTI	Department of Trade and Industry
EFB	empty fruit bunch
EIA	U.S. Energy Information Administration
EPA	U.S. Environmental Protection Agency

EtOH	ethanol
FC	fixed carbon
FIB	focused ion beam
FP	formulated pellets
FPU	filter paper unit
ft.	feet
g/cm^3	grams per cubic centimeter
GA	genetic algorithm
H	hydrogen
H_2O	water
HHV	high heat value
HMPP	high-moisture pelleting process
HPB	hydrophobic
HPL	hydrophyllic
HTC	hydrothermal carbonization
HTTP	hydrothermal pellets
IEA	International Energy Agency
IL	ionic liquid
in.	inch
INL	Idaho National Laboratory
ISO	International Organization for Standardization
ITEBE	French International Association of Bioenergy Professionals
K	potassium
kcal/kg	kilocalorie per kilogram
kg	kilogram
kg/m^3	kilograms per cubic meter
kg/min	kilograms per minute
kJ/kg	kilojoule per kilogram
kWh/t	kilowatt-hours per ton
L/D	length to diameter ratio
lb/ft^3	pounds per cubic foot
LHV	lower heating value
Mg	magnesium
MJ/kg	megajoules per kilogram
mm	millimeter
MMBTU	million British thermal units per pound
MPa	megapascal
MPa/s	megapascals per second
MSS	milled sunflower seeds
MSW	municipal solid waste
N	nitrogen
Na	sodium
NaOH	sodium hydroxide
NMR	nuclear magnetic resonance

NOx	nitrogen oxides
O	oxygen
O&M	operation and maintenance
OAc	1-ethyl-3-methylimidazolium acetate [C2mim]
ORNL	Oak Ridge National Laboratory
P	phosphorus
PC	pyrolysis char
PDU	process demonstration unit
PFFP	plate and frame filter press
PFI	Pellet Fuels Institute
PMB	packed moving bed
ppm	parts per million
psi	pounds per square inch
PSO	particle swarm optimization
rpm	revolutions per minute
RSM	response surface model
S	sulfur
SA	simulated annealing
SAA	soaking in aqueous ammonia
SD	standard deviation
SEM	scanning electron microscopy
SEP	steam-exploded pellets
SHF	separate hydrolysis with fermentation
SOT	state of technology
SP	straw pellet
SRC	short rotation coppice
SSF	saccharification fermentation
SS-NMR	solid-state nuclear magnetic resonance
TEA	techno-economic analysis
Tg	glass transition
ton/hr	metric ton per hour
TOP	torrefied pellets
TWC	torrefied wood chip
U.S.	United States
USDA	U.S. Department of Agriculture
UK	United Kingdom
VM	volatile material
VOC	volatile organic compound
WP	softwood pellet
wt.% (d.b.)	weight percent on a dry basis
WV	water vapor
XRD	X-ray diffraction

Chapter 1
Bioenergy Feedstock Types and Properties

Abstract Various woody, herbaceous, and municipal solid waste (MSW) biomass feedstocks such as such as lodgepole pine, hybrid poplar, corn stover, other agricultural straws, switchgrass, miscanthus, and other forms of MSW are available for bioenergy applications. All of these have different feedstock physical and chemical compositions. The focus of this chapter is to understand the physical properties of these biomass feedstock sources, such as bulk density, chemical composition (i.e., proximate and ultimate composition, energy content, and biochemical composition), and calorific value. This chapter also discusses the biomass types based on ISO standards, as well as providing information regarding the various standard methods that are available for measuring the raw and densified biomass physical properties, chemical, and biochemical composition.

Keywords Biomass · Woody · Herbaceous · Fruit · Municipal solid waste · Physical properties · Chemical and biochemical composition · Calorific value

1.1 Introduction

1.1.1 Biomass

World energy consumption has grown by 39% in the last two decades and is expected to grow by more than 50% in the next two decades. In today's world, fossil-based energy is a significant contributor. For example, petroleum, coal, and natural gas are the primary sources of energy. The challenge with fossil-based energy is that these types of fuel sources are finite—and they are depleting every day. Another major challenge in using fossil-based energy has to do with greenhouse gas emissions, which contributes to an increase in the Earth's temperature. Biomass is one of the largest providers of domestic renewable energy (accounting for 47%), which supplies over 3% of total energy consumption in the United States. In his review on energy production from biomass, McKendry (2002) pointed out how using biomass to produce energy can play a vital role in helping the developed world to reduce the environmental impact of burning fossil fuels.

© Springer Nature Switzerland AG 2020

J. S. Tumuluru, *Biomass Densification*,
https://doi.org/10.1007/978-3-030-62888-8_1

The 2015 Paris Agreement resulted in a global-wide commitment to reduce annual emissions of greenhouse gases by 2020 to keep the temperature of the Earth below 2 °C above pre-industrial levels (Clémençon 2016). Producing more energy from renewable energy resources, such as biomass, can help to reach this goal. Biomass is one of the significant sources of energy for humanity and is presently estimated to contribute on the order of 10–14% of the world's energy supply. The term "biomass" is used for all organic materials, including algae, trees, crops, and other sources. Biomass is considered to be organic because energy from the sun is stored in chemical bonds, per research conducted by the U.S. Energy Information Administration (EIA 2018).

According to the 2016 Billion-Ton Report jointly published by the U.S. Department of Energy (DOE) and Oak Ridge National Laboratory (ORNL), more than 1.2 billion tons of biomass is currently produced in the United States annually, which could sustainably be used for energy production (DOE 2016a). This abundance of biomass comes from a variety of sources, such as agricultural resources like cereal crops and oilseeds, as well as from perennial crops to potentially be planted on available forest-based idle acres. Indeed, biomass produced in the United States is well positioned to make a considerable impact on renewable energy production in the future (DOE 2016b).

1.1.2 Various Biomass Types According to the International Standards Organization (ISO-17225-1: 2014)

1.1.2.1 Woody Biomass

Woody biomass comes from trees, bushes, and shrubs—including forest, plantation, and other virgin wood that might have been subjected to size reduction, debarking, drying, or wetting. The forest, plantation, and other virgin wood also include wood from forests, parks, gardens, and plantations, as well as from short-rotation forests and coppice. The by-products and residues from the wood processing industry are also classified in this group. These sources can be chemically treated (e.g., glued, painted, coated, lacquered, or otherwise treated wood) or untreated (e.g., residues from debarking, sawing, size reduction, shaping, and pressing), as long as they do not have heavy metals or other halogenated organic compounds, which can be caused by treating this material with preservatives and coatings. Used wood, which is considered as post-consumer or post-society-used wood waste, also gets categorized as woody biomass. Only the natural or just the mechanically processed wood with insignificant contamination during its use can be categorized as this type of biomass, which includes the contaminants not found in the wood in its natural state (e.g., pallets, transport cases, boxes, wood packages, cable reels, construction wood). Used wood should not contain any higher amounts of heavy metals than that found in virgin wood, or halogenated organic compounds because of treating this wood with preservatives or coatings. Blends and mixtures also refer to woody biomass. The mixing can be intentional, which is categorized as blends, or unintentional, known as mixtures.

1.1.2.2 Herbaceous Biomass

Herbaceous biomass comes from plants and non-woody material, which die at the end of the growing season. Material that comes directly from the field after a period of storage and may only have been subject to size-reduction and drying is included in this grouping. It covers herbaceous material from agricultural and horticultural fields and gardens and parks. By-products and residues from the agricultural and horticulture industry refer to any herbaceous biomass material that is left over after industrial handling and treatment. The examples of residues are sugar from sugar beets, barley malt residues from beer production, and raw vegetable residues from the food processing industry. Biomass blends refer to the intentional mixing of the various herbaceous biomasses, while mixtures refer to the unintentional mixing of these sources.

1.1.3 Fruit Biomass

Fruit biomass sources are those that derive from orchards and horticulture fruit from trees, bushes, and fruit from herbs (e.g., tomatoes and grapes). The by-products of fruit biomass sources are those leftover after industrial handling and treatment.

1.1.3.1 Aquatic Biomass

Hydrophytic plants or hydrophytes that are adapted to living in or part of aquatic environments are considered as aquatic biomass. Aquatic biomasses are divided into the following main species: algae, water hyacinth, lake, and seaweed.

1.1.3.2 Blends and Mixtures

In the case of blends, the biomass is intentionally mixed, and in the case of mixtures, the biomass is mixed unintentionally. The origin of these blends and mixtures should be described.

1.1.4 Converting Biomass

During energy production using biochemical and thermochemical conversions—such as digestion, gasification, pyrolysis, and incineration—various bonds in the biomass are broken, which help to release stored energy (Gumisiriza et al. 2017). However, the major challenge in doing this is processing the biomass energy efficiently and meeting the necessary specifications for different conversion pathways. Researchers around the world have been developing novel ways to process and convert biomass into energy efficiently. If biomass is handled efficiently and sustainably, it can have a huge impact on the emissions of greenhouse gas. Energy from biomass is produced in either a solid or a liquid form. In solid form, biomass is burned in power generation plants to produce biopower, whereas biochemical and thermochemical conversion routes are used to produce liquid fuels. Another significant impact of using large amounts of biomass for bioenergy production is that it can have a very positive effect on rural economies.

The interest in utilizing biomass in large-scale applications has been increasing consistently in the last two decades. The initial focus was to use small percentages of biomass to "cofire" it with coal due to its less expensive costs, low risk, and more efficient processes for generating electricity (Tumuluru et al. 2012). The term "cofiring" means burning two types of material in a boiler. According to DOE, the definition of "cofiring" means mixing biomass with fossil fuels in conventional power plants following the modification of existing boilers. Up to 15% of biomass cofiring with coal is recommended after making boiler modifications (Tumuluru et al. 2012). Burning higher percentages of biomass with coal can result in the loss of boiler efficiencies, which is due to the moisture and energy content properties in the biomass. Another major limitation of burning biomass at higher percentages with coal is the quality of the ash content. In general, the ash produced from burning coal is used in cement production, but the ash from biomass is not currently used in the cement industry.

During 1995–2005, first-generation biofuels—mainly ethanol—derived from corn grain, sugarcane, and vegetable oil feedstock became available in the United States, European Union, United Kingdom, Brazil, and many other countries around the world. Still, they were limited due to the food versus fuel controversy. Since 2005, second-generation biofuels that are produced using non-food, lignocellulosic material, such as municipal waste and wood chips, along with dedicated crops like switchgrass and miscanthus, have gained a lot of attention. Currently, the use of municipal solid waste (MSW) for the generation of biofuels is gaining a lot of interest. According to Liu et al. (2015), proper MSW management is crucial for public health. Swanson (2008) also states that with the increase in the economies of many countries, there is a higher demand for natural resources and consumer goods. This greater demand for consumer goods has resulted in the generation of high amounts of solid waste. Improperly managed solid waste can result in: (a) risk to human health and the environment by contaminating water streams; (b) attracting rodents and inset growth; and (c) flooding due to blocked drainage of canals, water streams, and other water bodies (EPA 2002). The EPA developed a report that includes information on MSW generation, recycling, and disposal. In 2013, 254 million tons of trash were generated and recycled, while 87 million tons of material were composted, which is equivalent to a 34.3% recycling rate. The major challenge in using MSW or other biomass feedstocks for bioenergy applications is the variability in its physical properties and chemical composition (EPA 2013; Tumuluru 2016; Tumuluru 2018a).

Some of the major challenges in using diverse biomass resource types and the conditions under which they are produced include the variability in their composition and physical properties limits, as well as their commercial-scale applications. Another major challenge is the reliable supply of high-quality biomass to biorefineries. Technical limitations of using biomass for biopower and biofuels include its high moisture content, irregular size and shape, low bulk density, higher organic volatile matter, and lower energy density, thereby resulting in lower heating value and higher concentrations of alkali and alkaline-earth elements (e.g., potassium [K], sodium [Na], calcium [Ca], magnesium [Mg]), and higher concentrations of phosphorous (P), chlorine (Cl), and water vapor (WV). The elastic and cohesive nature of this biomass results in bridging during storage and feeding, which has a negative influence on grindability. In addition, high moisture content and a fibrous nature increases grinding energy and produces irregular particle sizes and shapes of biomass (Tumuluru et al. 2012; Tumuluru 2018a; Tumuluru et al. 2011). However, there is increased interest in utilizing biomass for biofuels in Europe and in the United States due to: (a) improvements in biomass production and conversion technologies; (b) the production of food in surplus; and (c) climate change due to greenhouse gas emissions (Popp et al. 2014). Figure 1.1 shows the various types of biomass that are typically used for biofuels and biopower generation.

1.2 Biomass Properties

1.2.1 Proximate and Ultimate Composition

The quality of the biomass used for biofuels and biopower generation is defined based on its individual chemical and physical properties. McKendry (2002) indicated that there are various biomass physical, chemical, and energy properties that have an impact on the conversion process. In his studies on the approaches to predict biomass cofiring with pulverized coal, Belošević (2010) indicated that variability in biomass physical and chemical properties has a significant impact on the characteristics of feeding and conversion. Fuel analysis is based on volatile material (VM) content, ash, and moisture, while fixed carbon (FC)—determined by difference—is termed the proximate analysis of fuel. Elemental analysis of fuel, presented as carbon (C), nitrogen (N), hydrogen (H), oxygen (O), and sulfur (S), together with ash content, is termed the ultimate analysis of fuel. The typical proximate and ultimate compositions of woody and herbaceous biomasses are given in Table 1.1

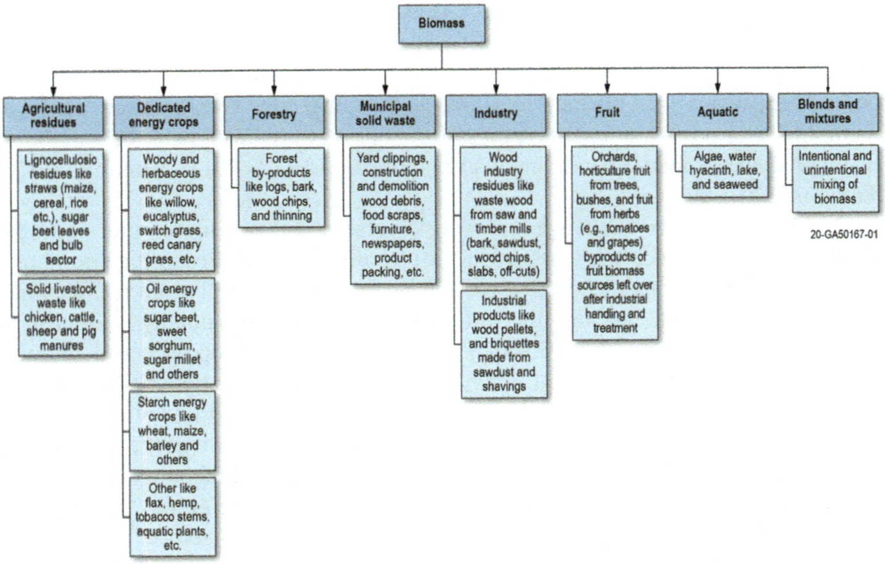

Fig. 1.1 Various types of biomass types available for bioenergy production

(Hartmann 2007). It is clear from Table 1.1 that woody biomass has a much higher moisture content as compared to herbaceous biomass. Woody biomass also has better fuel properties in terms of proximate and ultimate composition and energy properties as compared to herbaceous biomass. The content of VM and FC helps in understanding the ease with which the biomass can be burned and further gasified and oxidized. From Table 1.1, it is also clear that fossil fuels, such as coal, have higher C content but lower O content.

Proximate analysis of fuel indicates the percentage of the material that burns in a gaseous state and a solid state. For example, volatiles burn in a gaseous state, while fixed carbon (FC) and ash burn in a solid state. The proximate and ultimate analysis is developed for solid fuels, such as coal, which has energy stored within it in either a solid or a gaseous phase. VM and moisture content in the solid fuel is driven-off in the form of as-by heating at 950 °C for 7 min, whereas FC is the leftover mass after burning the volatiles. Tables 1.2 and 1.3 provide the chemical and ash composition of sorted MSW, which is used for gasification or incineration applications. Tables 1.4 and 1.5 provide the proximate composition of woody and herbaceous biomass sources, respectively, while Tables 1.6 and 1.7 provide the ultimate composition of woody and herbaceous biomass sources, respectively (Williams et al. 2017). Recent studies were conducted by Tumuluru and Fillerup (2020) on briquetting of blends of lodgepole pine, switchgrass, and corn stover. These authors have indicated that lodgepole pine, switchgrass, and corn stover have volatiles of about 85.46%, 79.53%, and 66.93% and fixed carbon values of about 13.73%, 15.05%, and 10.47%, respectively. The carbon, hydrogen, and nitrogen content of these

Table 1.1 Typical wood and herbaceous proximate and ultimate weight percent on a dry basis (wt. % [d.b.]) composition

	C	H	O	N	K	S	Cl
Spruce with bark	49.8	6.3	43.2	0.13	0.13	0.015	0.005
Beech with bark	47.9	6.2	43.3	0.22	0.22	0.015	0.006
Poplar short-rotation coppice (SRC)	47.5	6.2	44.1	0.42	0.35	0.031	0.004
Willow SRC	47.1	6.1	44.2	0.54	0.26	0.045	0.004
Coniferous bark	51.4	5.7	38.7	0.48	0.24	0.058	0.019
Virgin wood materials (coniferous wood)	47–54	5.6–7.0	40–44	<0.1–0.5	–	<0.01–0.05	<0.01–0.03
Virgin wood materials (deciduous wood)	48–52	5.9–6.5	41–45	<0.1–0.5	–	<0.01–0.05	<0.01–0.03
Virgin bark materials	51–56	5.9–6.5	36–43	0.3–1.2	–	0.02–0.20	<0.01–0.05
Virgin logging residues	50–53	5.9–6.3	40–44	0.3–0.8	–	0.01–0.08	<0.01–0.04
Virgin SRC	47–51	5.8–6.7	40–46	0.2–0.8	–	0.02–0.10	<0.01–0.05
Miscanthus	47.5	6.2	41.7	0.73	0.70	0.150	0.220
Wheat straw	45.6	5.8	42.4	0.48	1.00	0.082	0.190
Triticale (grains)	43.5	6.4	46.4	1.68	0.60	0.110	0.070
Rape cake	51.5	7.4	30.1	4.97	1.60	0.550	0.019
Fossil fuels							
Coal	72.5	5.6	11.0	1.30	–	0.940	<0.1
Lignite	65.9	4.6	23.0	0.70	–	0.390	<0.1
Heating oil	85–86	11–13	1–4	–	–	–	–
Natural gas	75	25	–	–	–	–	–

Derived from Hartmann (2007)

Table 1.2 Proximate composition of MSW

Feedstock	Moisture on a percent wet basis (%, w.b.)	Standard deviation (SD)	Ash (%)	FC (%)	SD
MSW	3.51	0.01	13.89	14.75	0.30

Table 1.3 Ultimate composition and higher heating value of MSW

Feedstock	C (%)	SD.	H (%)	SD.	N (%)	SD.	Higher heating value (Btu/lb)	SD
MSW	42.34	0.32	5.87	0.04	3.57	0.06	7608.48	160.71

feedstocks were also analyzed. Corn stover has carbon, hydrogen, and nitrogen content of 38.64%, 4.77%, and 0.66%; switchgrass has about 46.84%, 5.64%, and 0.64%; and lodgepole pine has about 50.48%, 5.88%, and 0.12%, respectively.

Table 1.4 Proximate feedstock compositions for specific herbaceous biomass; average (standard deviation)

Feedstock composition	Corn stover	Switchgrass	Sorghum	Energy cane (bagasse)	Mixed grasses	Miscanthus
Volatiles (%)	78.1 (5.0)	82.4 (4.1)	77.0 (3.7)	82.2 (1.9)	78.6 (2.8)	82.5 (3.5)
Ash (%)	6.3 (3.5)	4.0 (2.0)	7.2 (2.6)	3.4 (1.6)	6.6 (1.7)	2.6 (1.3)
FC (%)	15.6 (4.4)	13.6 (3.0)	15.7 (2.3)	14.4 (1.0)	14.8 (2.4)	14.8 (2.9)

Table 1.5 Ultimate feedstock compositions for specific woody feedstocks: average (standard deviation)

Feedstock composition	Shrub willow	Hybrid poplar	Pine	Other softwoods	Other hardwoods
H (%)	6.0 (0.2)	6.0 (0.1)	6.1 (0.1)	6.1 (0.1)	6.1 (0.1)
C (%)	50.3 (0.9)	50.0 (1.1)	51.5 (1.0)	51.8 (0.9)	50.2 (0.5)
N (%)	0.36 (0.10)	0.35 (0.17)	0.17 (0.12)	0.27 (0.21)	0.55 (0.49)
O (%)	42.6 (0.4)	42.8 (1.2)	41.4 (1.0)	39.7 (1.8)	41.1 (1.6)
S (%)	0.04 (0.01)	0.03 (0.01)	0.02 (0.01)	0.03 (0.01)	0.05 (0.05)

Table 1.6 Ultimate feedstock compositions for specific herbaceous biomass: average (standard deviation)

Feedstock composition	Corn stover	Switchgrass	Sorghum	Energy cane (bagasse)	Mixed grasses	Miscanthus
H (%)	5.7 (0.3)	5.9 (0.2)	5.7 (0.2)	6.1 (0.1)	5.8 (0.3)	5.8 (0.1)
C (%)	47.1 (2.3)	47.1 (1.1)	46.4 (1.3)	48.8 (0.9)	47.6 (1.1)	48.9 (1.5)
N (%)	0.63 (0.32)	0.60 (0.26)	1.04 (0.38)	0.43 (0.20)	1.38 (0.54)	0.35 (0.17)
O (%)	40.3 (2.2)	42.4 (2.3)	40.3 (0.6)	–	39.5 (0.7)	42.3 (1.1)
S (%)	0.14 (0.53)	0.06 (0.03)	0.11 (0.01)	–	0.12 (0.02)	0.04 (0.02)

Table 1.7 Proximate feedstock compositions for specific woody feedstocks: average (standard deviation)

Feedstock composition	Shrub willow	Hybrid poplar	Pine	Other softwoods	Other hardwoods
Volatiles (%)	84.7 (0.8)	84.0 (1.3)	83.5 (2.5)	81.3 (2.9)	85.1 (3.0)
Ash (%)	1.5 (0.4)	1.3 (0.5)	0.7 (0.6)	2.1 (2.0)	1.8 (1.2)
FC (%)	13.8 (0.7)	14.6 (0.1)	15.7 (1.9)	16.5 (1.6)	13.1 (1.8)

1.2.2 Ash Content

Biomass consists of inorganic matter called ash. There are two types of ash: introduced ash and anatomical ash. The total ash content, which is the sum of the introduced and anatomical ash, is dependent on the type of feedstock, harvest method followed, and soil condition in which it was harvested. Tumuluru et al.

Table 1.8 Multi-pass corn stover collected from Nebraska in 2010, focusing on residue removal rates

Windrower	Removal	Ash (wt.% d.b.)
Bar rake	Low	12.9
Bar rake	High	14.5
Wheel rake	Low	8.8
Wheel rake	High	11.5
Flail shredder	Low	8.1
Flail shredder	High	9.3
Flail shredder	High	10.9

Derived from Tumuluru et al. (2016)

(2016) indicated that ash content decreases with maturity. These authors identified some of the factors that can influence the ash content and composition, such as delaying the harvest from fall to late winter results in environmental changes, soil type, soil chemistry, water quality, fertilization chemistry, and elemental composition of the soil. Table 1.8 indicates how residual removal rates impact the ash content of corn stover.

Following the biochemical and thermochemical conversion of the biomass, the leftover ash is called the solid residue. In the case of thermochemical conversion, higher ash content presents significant operational and conversion problems. The ash can form slag at higher temperatures, which reduces a plant's operating capacity and increases costs. In the case of biochemical conversion, the nonhydrolyzed material left over after enzymatic conversion is a combination of the ash and remaining C, which is not converted further. The other impact that higher ash content in biomass can have is in the handling and processing costs. The variability in terms of chemical and ash composition can have a significant impact on the overall economics of biofuels production. Anatomical ash in biomass has an alkali metal content, such as Na, K, Mg, P, and Ca, which are important during thermochemical conversion. At higher temperatures, alkali metals with silica can produce slag, which is sticky and mobile in a liquid phase and can block furnaces and boiler air pathways. Recent studies at Idaho National Laboratory (INL) on processing biomass with higher ash content have indicated that the wear and tear of the preprocessing equipment is higher. This can result in the frequent changeout of processing equipment parts, which can have a significant impact on processing costs. Recent interactions with the preprocessing industry have suggested that when biomass material with a higher ash content is hammermilled, the need exists to change the hammer every week. Table 1.9 shows the ash content and ash melting temperature of different types of woody and herbaceous biomass.

1.2.3 Calorific Value

Table 1.10 provides the typical calorific values (CVs) of woody and herbaceous biomass. The energy content is measured as heat content in the sample per unit mass

Table 1.9 Ash content and ash melting temperature of different woody, herbaceous, and municipal solid waste biomass

Feedstock type	Ash (wt.% d.b.)
Typical calorific values (CVs) for virgin wood materials (coniferous wood)	0.2–0.5
Typical CVs for virgin wood materials (deciduous wood)	0.2–0.5
Typical CVs for virgin bark materials	2–10
Typical CVs for virgin logging residues	1.5–2
Typical CVs for willow and poplar SRC	2
Spruce with bark	0.6
Beech with bark	0.5
Poplar (SRC)	1.8
Willow (SRC)	2.0
Coniferous tree bark	3.8
Vinewood chips	3.4
Miscanthus	3.9
Wheat straw	5.7
Triticale (grains)	2.1
Rape cake	6.2
Municipal solid waste (INL experimental data)	3.74
Corn stover	22.60
Switchgrass	5.42
Lodgepole pine	0.8

Derived from Francescato et al. (2008), Tumuluru and Fillerup (2020)

or volume (McKendry 2002). The CV is usually measured using an adiabatic or isoperibolic calorimeter. The heat content of the biomass is dependent on the C, H, and O content in the biomass. CV is the amount of heat content in biomass when burned in air. CV is normally expressed as either a higher heating value or a lower heating value. Higher heating value is the total energy in the material when burned in the air (this calculation also includes latent heat of vapor). In a practical application, the water vapor in the material cannot be used effectively; therefore, a lower heating value is more appropriate to use for the energy calculation. CV is usually determined by a calorimeter, but a theoretical value can be calculated from Eq. (1.1):

$$\text{Calorific value (kcal/kg of fuel)} = (8100\text{C} + 34000(\text{H-0/8}))/100 \tag{1.1}$$

where C, H, and O represent the percentages in 1 kg of fuel.

C yields 8080 kcal/kg when completely burned, H yields 34,000 kcal/kg, and O is already assumed to be attached to its proportion of H; that is, an amount of H equal to one-eighth the weight of the O is nullified. The S compounds are assumed to have their combustion heat nullified by the O-N compounds (Woodyard 2009). The calorific values of some herbaceous and woody biomasses are given in Table 1.10. Many researchers have modeled the CV calculation using the ultimate composition data. Velázquez-Martí et al. (2017) modeled the CV of biomass from fruit trees using elemental analysis data. These authors have developed indirect models for heating

Table 1.10 Calorific value of specific woody and herbaceous biomasses measured

Feedstock type	Net CV (M.J./kg)
Typical CVs for virgin wood materials (coniferous wood)	18.8–19.2
Typical CVs for virgin wood materials (deciduous wood)	18.5–19.2
Typical CVs for virgin bark materials	19–21
Typical CVs for virgin logging residues	19–20
Typical CVs for short-rotation coppice (SRC) (willow and poplar)	18.6–19.2
Spruce with bark	18.8
Beech with bark	18.4
Poplar (SRC)	18.5
Willow (SRC)	18.4
Coniferous tree bark	19.2
Vinewood chips	19.8
Miscanthus	17.6
Wheat straw	17.2
Triticale (grains)	16.9
Rape cake	21.2
Lodgepole pine	20.24
Switchgrass	19.19
Corn stover	15.90

Derived from Francescato et al. (2008), Tumuluru and Fillerup (2020)

value prediction of biomass from orange trees (*Citrus* × *sinensis Osbeck*), almond trees (*Prunus dulcis* [Mill]), and olive trees (*Olea europaea L*) using elemental analysis data. They reasoned that the development of these equations reduces: (a) the biomass CV determination time; and (b) the biomass characterization costs. The equations developed by these authors for the CV calculation are given in Table 1.11.

1.2.4 Biochemical Composition

The typical biochemical compositions (e.g., lignin, protein, extractives) of specific herbaceous and woody biomasses are given in Tables 1.12, 1.13, and 1.14 (Williams et al. 2017). In general, woody biomass has a higher lignin content as compared to herbaceous biomass. The amount of lignin and cellulose in the biomass is important in the biochemical conversion process. In the case of biochemical conversion, the feedstock that has a higher percentage of cellulose and hemicellulose is preferred. The lignin in the biomass is potentially an abundant energy source, which is left over after hydrolysis. This lignin is currently used for burning in boilers. Many researchers are working on utilizing lignin for producing higher-value products.

Table 1.11 Equations developed for calculating higher heating values of various biomass species

Species	Model	R^2 adj. (%)
Mixing three species	$HHV = 717.79 + 380.85C + 7.61\ \%\ leaves - 14.00\ \%\ bark$	97.35
	$HHV = 863.61 + 383.93C - 15.257\ \%\ leaves$	96.97
	$HHV = 769.58 + 371.80C + 8.02\ \%\ leaves$	97.14
	$HHV = 928.66 + 374.30C$	96.73
Prunus dulcis	$HHV = 7624.11 + 237.76C + 228.75N - 83.37\ \%\ w$	97.13
	$HHV = -1024.04 + 412.99C + 272.54N$	96.74
	$HHV = -1106.84 + 422.77C$	94.63
Citrus × sinensis	$HHV = 4323.54 + 301.46C - 72.74\ \%\ bark$	88.12
	$HHV = 1254.39 + 365.79C$	83.17
Olea europaea	$HHV = -790.92 + 421.37C$	94.85

Derived from Velázquez-Martí et al. (2017)
HHV High heat value (kJ/kg), *C* Carbon, *N* Nitrogen, %w: Moisture content in wet basis

Table 1.12 Average biochemical composition of specific herbaceous feedstocks (standard deviation)

Feedstock composition	Corn stover	Switchgrass	Sorghum	Energy cane (bagasse)	Mixed grasses	Miscanthus
Cellulose (%)	34.3 (2.5)	34.2 (2.7)	28.6 (2.6)	32.1 (3.2)	28.9 (2.9)	38.9 (3.2)
Hemicellulose (%)	20.7 (2.0)	21.9 (2.6)	15.4 (1.6)	19.5 (1.9)	16.7 (3.9)	20.1 (1.4)
Lignin (%)	15.2 (1.6)	19.2 (1.4)	12.2 (1.9)	16.3 (1.8)	15.7 (1.7)	21.1 (1.6)

Table 1.13 Average biochemical composition of specific woody biomasses (standard deviation)

Feedstock composition	Hybrid poplar	Pine	Other softwoods	Other hardwoods
Cellulose (%)	43.8 (1.2)	47.4 (2.2)	42.1 (7.1)	50.8 (6.9)
Hemicellulose (%)	14.7 (0.1)	21.9 (4.9)	25.1 (5.2)	29.7 (4.3)
Lignin (%)	25.7 (0.3)	28.6 (0.7)	29.1 (1.7)	19.5 (4.1)

Table 1.14 Biochemical composition of woody and herbaceous biomass

Feedstock type	% Total protein	% Lignin	% Extractives	
			% Ethanol (EtOH)	% Water (H_2O)
Corn stover	2.54	15.29	2.64	6.52
Wheat straw	3.59	13.19	2.41	7.34
Lodgepole pine	0.21	26.48	4.43	4.06
Pinyon-juniper	1.83	32.86	5.40	7.65
Sorghum	4.03	12.90	2.89	13.49
MSW (INL data)	0.97	25.79	1.38	4.08

Derived from Tumuluru (2018b)

1.2.5 Moisture

Table 1.15 shows the moisture content of woody and herbaceous biomass. Both woody and herbaceous biomass is available at higher moisture content. The herbaceous biomass moisture content depends on the type of harvesting method followed and the harvesting time. The water in the biomass is present in two forms—bound and free water. Bound moisture is present in the cell walls, whereas the free water is available in liquid pockets in the void spaces. In general, the transition zone where the bound and free moisture occurs is called the fiber saturation point (Stamm 1971). Normally, biomass is dried below the fiber saturation point to keep it aerobically stable. Various active drying technologies are used to dry the biomass to keep it aerobically stable. For example, a rotary dryer is used to dry large quantities of woody and herbaceous biomass. However, the two major challenges in using rotary dryers are the capital and operating costs (Lamers et al. 2015). These dryers operate at high temperatures (i.e., generally, the inlet temperature is about 500 °C, while the outlet temperature is kept at about 212 °C to avoid condensation of the moisture as the material exits the dryer). In addition, high-temperature drying results in volatile organic compound (VOC) emissions, which are not good for the environment, nor human beings should they happen to inhale it (Tumuluru 2016). Low-temperature dryers, such as belt dryers, operate between 60 °C and 80 °C and are now being widely used in wood chip drying. These dryers require less capital and operating costs and can typically avoid the environmental concerns frequently observed in high-temperature dryers (Tumuluru 2016). Moisture content in biomass has a great impact on biomass processing, especially size reduction. In their studies on using different types of grinders (e.g., hammermill, rotary shear, vortex mill), Tumuluru and Yancey (2018) indicated that increasing moisture content increases grinder energy consumption and has a great impact on particle size distribution.

1.2.6 Bulk Density

Bulk density of biomass is an important characteristic as it has a great impact on feeding, storage, and transportation. Woody biomass has a higher density as compared to herbaceous biomass, as shown in Table 1.16. Tumuluru et al. (2015) discussed the physical properties of woody and herbaceous biomass and thermally

Table 1.15 Typical moisture content of woody and herbaceous biomass after harvesting

Feedstock type	Moisture content (%, w.b.)
Corn stover after harvesting	5–50% (w.b.) (Tumuluru et al. 2016; Womac et al. 2005)
Stem wood at harvest	45–50% (w.b.) (Saralecos et al. 2014)
Energy sorghum stems at harvest	70–85% (w.b.) (McKinley et al. 2018)
Switchgrass after harvest	70% (w.b.) (Khanchi and Birrell 2015)
Small diameter fuel wood	>50% (w.b.) (Routa et al. 2016)

Table 1.16 Bulk density of woody and herbaceous biomass

Feedstock type	Bulk density
Corn stover	157.9 (3.27)
Wheat straw	151.8 (3.85)
Lodgepole pine	246.2 (4.2)
Pinyon-juniper	257.47 (7.1)
Sorghum	223.7 (1.46)

Derived from Tumuluru (2018b)

Table 1.17 Physical properties of herbaceous, woody, and torrefied biomass ($n = 3$)

Feedstock type	Moisture content (%, w.b.)	Bulk density (kg/m^3)	Particle density (kg/m^3)	Porosity
Ground switchgrass	10.21	151	650	0.76
Wood pellets	5.02	710	1210	0.41
Wood chips	12.12	265	580	0.54
Torrefied wood chips	1.8	225	505	0.55

Table 1.18 Ground MSW physical properties from a stage-1 grinder

	Particle size (mm)					Physical properties	
Stage-1 Grinder screen size	Geometric mean particle size	SD	d10 (mm)	d50 (mm)	d90 (mm)	Bulk density (kg/m^3)	SD
50.8	6.16	1.02	0.83	7.29	11.53	25.34	3.88

Table 1.19 Ground MSW physical properties from a stage-2 grinder

	Particle size (mm)					Physical properties	
Stage-2 Grinder screen size	Geometric mean particle size	SD	d10 (mm)	d50 (mm)	d90 (mm)	Bulk density (kg/m^3)	SD
19.05	3.62	0.36	0.24	3.34	8.79	32.88	1.64
12.7	2.35	0.21	0.17	2.50	10.93	40.06	4.32
6.35	1.98	0.98	0.20	1.78	5.45	57.74	20.82

pretreated biomass used for biofuels production, as shown in Table 1.17. It is clear from Table 1.17 that the bulk density of woody biomass increases by almost four to five times after densification using a pellet mill. The porosity of the biomass decreases after densification. Thermally pretreated biomass has a lower bulk density as compared to raw woody biomass. This is primarily due to the loss of moisture and volatiles after thermal pretreatment. The bulk density and particle dimensions—such as d10, d50, d90, and geometric mean particle length for an MSW grind in a hammermill (stage-1 and stage-2 grinder) fitted with different screens—are given in Tables 1.18 and 1.19, respectively. The measured particle dimensions are greatly impacted by screen size. In addition, the MSW bulk density changed with the screen size of the grinder. Smaller grinder screen sizes increase the bulk density marginally.

In his studies, Tumuluru (2019) found that the bulk density of the raw corn stover, switchgrass, and lodgepole pine increased after briquetting. Lodgepole pine briquettes recorded a maximum bulk density of 489 kg/m^3, switchgrass recorded about 446 kg/m^3, and corn stover recorded about 497 kg/m^3. They also reported that lodgepole pine when ground using a hammermill fitted with a 4.8 and a 12.7 mm screen size produced particle with geometric mean particle length of about 1.13 and 1.64 mm; switchgrass recorded about 0.75 and 2.35; and corn stover measured about 0.85 and 1.05 mm, respectively.

To increase the density of herbaceous biomass, they are baled and transported. The density of woody chips and bales do not meet transportation efficiencies. To further increase transportation efficiencies, biomass is densified using mechanical compression systems. Commonly used densification systems include the pellet mill, briquette press, and cubers. When compressed and extruded, the bulk density of these systems increases both the woody and herbaceous biomasses by almost four to five times. After using these systems, the final bulk densities of woody and herbaceous biomass will be in the range of 450–700 kg/m^3. The commonly used standards for measuring the raw and densified biomass physical, chemical, and biochemical compositions are given in Table 1.20. There are ISO (2014a), ASTM (2015a, b, 2019), ASABE (1988, 2008), and PFI methods for physical properties and chemical composition, and National Renewable Energy Laboratory (NREL), Laboratory Analytical Procedure (LAP), and National Institute of Standards and Technology (NIST) methods for biochemical composition.

1.2.7 Biomass Limitations

Harvested biomass lacks both the physical properties and chemical composition to produce biofuels on a commercial scale. The physical properties, such as low bulk density, irregular size, and inconsistent shape, result in feed, handling, and transportation problems. Flow issues associated with biomass also make it difficult to move from one location to another. Limitations in the physical properties result in logistics issues as well. The lower-energy content of biomass fuels results in high volumes of biomass requirements, which again result in storage, transportation, and feed handling issues for both biochemical and thermochemical conversions and biopower generation.

The moisture variability in biomass is another major limitation for processing and conversion to fuels. High moisture in biomass creates physical and chemical properties issues. Preprocessing, feeding, and handling of biomass are influenced by moisture variability. In preprocessing, moisture in the biomass influences the grinding process. Higher moisture content results in higher grinding energy and produces variable particle sizes. In their studies on the grinding energy and physical properties of chopped and hammer-milled barley, wheat, oat, and canola straws, Tumuluru et al. (2014) indicated that the fibrous nature of biomass increases with moisture content and increases the grinding energy. In addition, higher moisture content

Table 1.20 Standards commonly used for physical properties, chemical, and biochemical composition, and calorific value of raw and densified biomass

ISO standard references for biomass	
CEN/TS 14779 (CEN/TS 2005a)	Solid biofuels—sampling—methods for preparing sampling plans and sampling certificates
CEN/TS 14780 (CEN/TS 2005b)	Solid biofuels—methods for sample preparation
EN 14774-3 (EN 2009a)	Solid biofuels—determination of moisture content—Stufe drying method part 3 moisture content analysis for overall sample analysis
EN 14918 (EN 2009b)	Solid biofuels—determination of calorific value
CEN/TS 15104 (CEN/TS 2005c)	Solid biofuels—determination of total carbon, hydrogen, and nitrogen—instrumental methods
CEN/TS 15290 (CEN/TS 2006)	Solid biofuels—determination of major elements
EN 14961-1 (EN 2009c)	Solid biofuels—fuel specifications and classes—part 1: general requirements
ISO 17225-1:2014b (ISO 2014b)	Solid biofuels—fuel specifications and classes—part 1: general requirements
ISO 17830:2016 (ISO 2016)	Solid biofuels—particle size distribution of disintegrated pellets
ISO 17830:2016 (ISO 2016)	Determination of particle size distribution of material within pellets
ISO 18125:2017 (ISO 2017)	Solid biofuels—determination of calorific value
ISO 16559:2014c(en) (ISO 2014c)	Solid biofuels—terminology, definitions, and descriptions
American Society of Agricultural and Biological Engineers standards for biomass physical properties	
American Society of Agricultural and Biological Engineers society has developed standards for physical properties such as moisture content, density, particle size distribution, and durability of raw and densified products	ASABE S269.4, Cubes, Pellets, and Crumbles—Definitions and Methods for Determining Density, Durability, and Moisture Content (ASABE 2008)
	ASABE S358.2, Moisture Measurement—Forages (ASABE 1988)
	ANSI/ASAE S424.1, Method of Determining and Expressing Particle Size of Chopped Forage Materials (ANSI/ASAE 1992)
ASTM International Standards for proximate and ultimate composition and calorific value	
ASTM standards are commonly used for proximate and ultimate composition analysis and calorific value	ASTM D7582–15, Standard Test Methods for Proximate Analysis (ASTM 2015b)
	ASTM D3176–15, Standard Practice for Ultimate Analysis (ASTM 2015a)

(continued)

Table 1.20 (continued)

ISO standard references for biomass	
	ASTM D5865/D5865M-19, Standard Test Method for Gross Calorific Value (ASTM 2019)
Pellet Fuel Institute	
Pellet Fuel Institute (PFI) has standards for pellets to be traded in the United States	The PFI standards program is a third-party accredited program that enables consumers to easily identify PFI graded fuel—consistent, reliable pellets from producers whose facilities are subject to regular third-party inspection and lab testing.
NREL and NIST standards for biochemical composition	
The biochemical composition, such as lignin, hemicellulose, and cellulose, is measured using NREL methods (Sluiter et al. 2012)	
%Structural ash is measured from the extractive-free solid and is corrected to a whole mass basis using the %Total extractives value	
%Extractable inorganics is calculated by subtracting %Structural ash from %Whole ash	
%Extractable protein is calculated by subtracting %Structural protein from %Whole protein	
%H_2O extractives others is calculated by subtracting the components listed above as being a part of %H_2O extractives from the measure H_2O extractives mass value	
There are two different methods to analyze non-structural sugars: one includes %sucrose and the other includes %free glucose using a megazyme kit	
National Institute of Standards and Technology, US Department of Commerce (NIST), methods are also used for measurement of the chemical and biochemical composition of the biomass.	

makes biomass more elastic and cohesive, which in turn increases the feed handling, plugging of screens, and bridging of particles in feedstock bins.

First-generation biorefineries are having challenges in feeding, handling, transportation, and storage of raw biomass with variable physical properties, such as variable moisture and formats and different physical properties and chemical compositions. These properties result in biomass bridging with uneven flows into the bioreactor, thereby impacting the efficiency of the reactor. The physical, chemical, and rheological properties of raw biomass are currently limiting biorefineries to operate at their desired capacities. Solving these problems can increase production capacity, reduce the interruption of normal operations, and reduce preprocessing and conversion costs. Solving the biomass inherent physical, chemical, and rheological issues is crucial to enable biorefineries to operate at their designed capacities.

Preprocessing of biomass using mechanical systems, such as grinding and densification, helps to improve the biomass bulk density limitations, size, shape, and moisture variability issues. Improving these specifications help to improve the front-end unit operations of the biorefineries. The major limitation of using current mechanical processing systems for biomass processing is their high costs. The INL studies indicate that preprocessing costs are the major bottleneck. The industry is currently looking at developing novel energy-efficient grinding and densification technologies. In general, the grinding energies and particle size distribution of biomass are primarily influenced by moisture content. Higher moisture content results in higher grinding energy and uneven particle size distribution. Commonly used grinding systems, such as hammermills, are not suitable for high moisture biomass. New grinding systems, such as the rotary shear developed by the Forest Concepts company (Tumuluru and Yancey 2018), are less sensitive to moisture content and are gaining attention. The major limitation in using densification for biomass is that it has to be dried to about 10% (w.b.) moisture content before it can be compressed and extruded in a pellet mill or briquette press. Optimizing the grinding and densification systems for various biomass feedstock properties is very important to reduce preprocessing costs.

1.3 Conclusions

The major limitation of using biomass for biofuels application at the commercial scale is its variable physical properties and different chemical compositions. Woody and herbaceous biomass has different physical properties and chemical and biochemical compositions. Woody biomass has high bulk densities as compared to herbaceous biomass. In addition, woody biomass, when harvested, has a higher moisture content as compared to herbaceous biomass. Typical woody biomass moisture content after harvest is around 50–60% (w.b), whereas herbaceous biomass is around 20–30% (w.b.). In terms of the chemical composition, woody biomass has a higher C content and a lesser ash content. In contrast, herbaceous biomass has a lesser C content and a higher ash content. Also, biomass physical property limitations result in logistics, feeding, handling, storage, and transportation issues. Various preprocessing and pre-treatments can help to improve biomass physical properties and chemical compositions. Mechanical preprocessing systems, such as grinding and densification, are commonly used techniques that can help to improve biomass physical properties like density, size, and shape, and reduce the overall moisture content. Typical densification systems used for biomass are the pellet mill and briquette press. These systems help to increase the bulk density of biomass by four to five times. Optimizing the grinding and densification systems to meet the desired feedstock properties is very important to make biomass easy to store, transport, and feed.

References

ANSI/ASAE S424.1 (1992) Method of determining and expressing particle size of chopped forage materials by screening. American Society of Agricultural and Biological Engineers, March 1992

ASABE S358.2 (1988) Moisture measurement—Forages. American Society of Agricultural and Biological Engineers, St. Joseph, MI, USA

ASABE S269.4 (2008) Cubes, pellets, and crumbles—Definitions and methods for determining density, durability, and moisture content. American Society of Agricultural and Biological Engineers, St. Joseph, MI, USA

ASTM D3176-15 (2015a) Standard practice for ultimate analysis of coal and coke. American Society for Testing and Materials International, West Conshohocken, PA, USA
ASTM D7582-15 (2015b) Standard test methods for proximate analysis of coal and coke by macro thermogravimetric analysis. American Society for Testing and Materials International, West Conshohocken, PA, USA
ASTM D5865/D5865M-19 (2019) Standard test method for gross calorific value of coal and coke. American Society for Testing and Materials International, West Conshohocken, PA, USA
Belošević S (2010) Modeling approaches to predict biomass cofiring with pulverized coal. Open Thermodyn J 4:50–70
CEN/TS 14779 (2005a) Solid biofuels – Sampling – Methods for preparing sampling plans and sampling certificates. BSI British Standards, London, England
CEN/TS 14780 (2005b) Solid biofuels. Methods for sample preparation. BSI British Standards, London, England
CEN/TS 15104 (2005c) Solid biofuels. Determination of total content of carbon, hydrogen and nitrogen. Instrumental methods. BSI British Standards, London, England
CEN/TS 15290 (2006) Solid biofuels. Determination of major elements. BSI British Standards, London, England
Clémençon R (2016) The two sides of the Paris climate agreement: dismal failure or historic breakthrough? J Environ Dev 25(1):3–25
DOE (2016a) Billion-ton report: Advancing domestic resources for a thriving bioeconomy, Volume 1: Economic availability of feedstocks, ORNL/TM-2016/160. Oak Ridge National Laboratory, Oak Ridge. 448p. https://doi.org/10.2172/1271651
DOE (2016b) Billion-ton report: Advancing domestic resources for a thriving bioeconomy, Volume 2: Environmental sustainability effects of select scenarios from Volume 1, ORNL/TM-2016/727. Oak Ridge National Laboratory, Oak Ridge. 642p. https://doi.org/10.2172/1338837
EIA (2018) Biomass explained, 21 June 2018. https://www.eia.gov/energyexplained/biomass/. Accessed 5 Mar 2020
EN 14774-3 (2009a) Solid biofuels. Determination of moisture content. Oven dry method. Moisture in general analysis sample. BSI British Standards, London, England
EN 14918 (2009b) Solid biofuels. Determination of calorific value. BSI British Standards, London, England
EN 14961-1 (2009c) Solid biofuels-fuel specifications and classes, Part 6- non-woody pellets for non-industrial use. BSI British Standards, London, England
EPA (2002) Protecting water resources with smart growth. EPA 231-R-04-002, May 2004. Washington DC, USA. https://www.epa.gov/sites/production/files/2014-04/documents/protecting-water-resources.pdf. Accessed 5 Mar 2020
EPA (2013) Municipal solid waste, 29 March 2016. https://archive.epa.gov/epawaste/nonhaz/municipal/web/html/. Accessed 5 Mar 2020
Francescato V, Antonini E, Bergomi LZ, Metschina C, Schnedl C, Krajnc N, Koscik K, Gradziuk P, Nocentini G, Stranieri S (2008) Wood fuels handbook: production, quality requirements, trading. AIEL – Italian Agriforestry Energy Association, Agripolis, Italy. http://www.biomasstradecentre2.eu/scripts/download.php?file=/data/pdf_vsebine/literature/wood_fuels_handbook.pdf. Accessed 5 Mar 2020
Gumisiriza R, Hawumba JF, Okure M, Hensel O (2017) Biomass waste-to-energy valorisation technologies: A review case for banana processing in Uganda. Biotechnol Biofuels 10(11):1–29
Hartmann H (ed) (2007) Handbook bioenergy small plants. 2nd ed. Special publication by the German Federal Ministry of Food, Agriculture, and Consumer Protection (BMVEL) and the German Agency for Renewable Resources (FNR), Gülzow, 224 S
ISO (2014a) Solid biofuels — Fuel specifications and classes — Part 1: General requirements. ISO-17225-1:2014, May 2014
ISO 17225-1 (2014b) Solid biofuels — Fuel specifications and classes — Part 1: General requirements. BSI British Standards, London, England
ISO 16559 (2014c) Solid biofuels — Terminology, definitions, and descriptions. BSI British Standards, London, England

ISO 17830 (2016) Solid biofuels — Particle size distribution of disintegrated pellets. BSI British Standards, London, England

ISO 18125 (2017) Solid biofuels — Determination of calorific value. BSI British Standards, London, England

Khanchi A, Birrell S (2015) Influence of weather and swath density on drying characteristics of corn stover and switchgrass. 2015 American Society of Agricultural and Biological Engineers (ASABE) Annual International Meeting, Paper Number 152190753, 26–29 July 2015, New Orleans, LA, USA

Lamers P, Roni MS, Tumuluru JS, Jacobson JJ, Cafferty KG, Hansen JK, Kenney K, Teymouri F, Bals B (2015) Techno-economic analysis of decentralized biomass processing depots. Bioresour Technol 194:205–213

Liu YY, van Dijk AIJM, de Jeu RAM, Canadell JG, McCabe MF, Evans JP, Wang G (2015) Recent reversal in loss of global terrestrial biomass. Nat Clim Change 5:470–474

McKendry P (2002) Energy production from biomass (Part 1): Overview of biomass. Bioresour Technol 83:37–46

McKinley BA, Olson SN, Ritter KB, Herb DW, Karlen SD, Lu F, Ralph J, Rooney WL, Mullet JE (2018) Variation in energy sorghum hybrid TX08001 biomass composition and lignin chemistry during development under irrigated and non-irrigated field conditions. PLoS One 13(4): e0195863

Popp J, Lakner Z, Harangi-Rákos M, Fári M (2014) The effect of bioenergy expansion: Food, energy, and environment. Renew Sust Energ Rev 32:559–578

Routa J, Kolström M, Ruotsalainen J, Sikanen L (2016) Validation of prediction models for estimating the moisture content of logging residues during storage. Biomass Bioenergy 94:85–93

Saralecos JD, Keefe RF, Tinkham WT, Brooks RH, Smith AMS, Johnson LR (2014) Effects of harvesting systems and bole moisture loss on weight scaling of Douglas-Fir sawlogs (*Pseudotsuga Menziesii* var. *glauca Franco*). Forests 5(9):2289–2306

Sluiter A, Hames B, Ruiz R, Scarlata C, Sluiter J, Templeton D, Crocker D (2012) Determination of structural carbohydrates and lignin in biomass. National Renewable Energy Laboratory (NREL) Laboratory Analytical Procedure (LAP), NREL/TP-510-42618

Stamm AJ (1971) Review of nine methods for determining the fiber saturation points of wood and wood products. Wood Sci 42:114–128

Swanson T (2008) Economic growth and environmental regulation: A discussion of international experiences. University College of London, London, England

Tumuluru JS (2019) Effect of moisture content and hammer mill screen size on the briquetting characteristics of woody and herbaceous biomass. KONA Powder Part J 36:241–251

Tumuluru JS (2016) Specific energy consumption and quality of wood pellets produced using high moisture lodgepole pine grind in a flat die pellet mill. Chem Eng Res Des 110:82–97

Tumuluru JS (2018a) Effect of pellet die diameter on density and durability of pellets made from high moisture woody and herbaceous biomass. Carbon Resour Convers 1(1):44–54

Tumuluru JS (2018b) Why biomass preprocessing and pretreatments? In: Tumuluru JS (ed) Biomass Preprocessing and Pretreatments for Production of Biofuels. CRC Press, Boca Raton, pp 1–14

Tumuluru JS, Fillerup E (2020) Briquetting characteristics of woody and herbaceous biomass blends: Impact on physical properties, chemical composition, and calorific value. Biofuels Bioprod Biorefin, First published: 03 July 2020. https://doi.org/10.1002/bbb.2121

Tumuluru JS, Hess JR, Boardman RD, Wright CT, Westover TL (2012) Formulation, pretreatment, and densification options to improve biomass specifications for cofiring high percentages with coal. Ind Biotechnol 8(3):113–132

Tumuluru JS, Lim CJ, Bi XT, Kuang X, Melin S, Yazdanpanah F, Sokhansanj S (2015) Analysis on storage off-gas emissions from woody, herbaceous, and torrefied biomass. Energies 8 (3):1745–1759

Tumuluru JS, Searcy E, Kenney KL, Smith WA, Gresham GL, Yancey NA (2016) Impact of feedstock supply systems unit operations on feedstock cost and quality for bioenergy applications. In: Kumar R, Singh S, Balan V (eds) Valorization of Lignocellulosic Biomass in a Biorefinery: From Logistic to Environmental and Performance Impact. Nova Science Publishers, New York, NY, USA, pp 1–45

Tumuluru JS, Tabil LG, Song Y, Iroba KL, Meda V (2014) Grinding energy and physical properties of chopped and hammer-milled barley, wheat, oat, and canola straws. Biomass Bioenergy 60:58–67

Tumuluru JS, Wright CT, Hess JR, Kenney KL (2011) A review of biomass densification systems to develop uniform feedstock commodities for bioenergy application. Biofuels Bioprod Biorefin 5(6):683–707

Tumuluru JS, Yancey N (2018) Conventional and advanced mechanical preprocessing methods for biomass: performance quality attributes and cost analysis. In: Tumuluru JS (ed) Biomass Preprocessing and Pretreatments for Production of Biofuels. CRC Press, Boca Raton, FL, USA, pp 15–49

Velázquez-Martí B, López-Cortés I, Salazar-Hernández D, Callejón-Ferre ÁJ (2017) Modeling the calorific value of biomass from fruit trees using elemental analysis data. In: Tumuluru JS (ed) Biomass Volume Estimation and Valorization for Energy. IntechOpen, Zagreb, pp 271–292

Williams CL, Emerson RM, Tumuluru JS (2017) Biomass compositional analysis for conversion to renewable fuels and chemicals. In: Tumuluru JS (ed) Biomass Volume Estimation and Valorization for Energy. IntechOpen, Zagreb, pp 251–270

Womac AR, Igathinathane C, Sokhansanj S, Pordesimo LO (2005) Biomass moisture relations of an agricultural field residue: Corn stover. Trans ASAE 48:2073–2083

Woodyard D (2009) Fuels and lubes: Chemistry and treatment. In: Woodyard D (ed) Pounders Marine Diesel Engines and Gas Turbines. Butterworth-Heinemann Publishers, Oxford, England, pp 88–141

Chapter 2
Binding Mechanism, Densification Systems, Process Variables, and Quality Attributes

Abstract Raw biomass is difficult to use as a feedstock for biofuels production at the commercial scale. In its raw form, biomass is bulky, wet, and scattered. Densification technologies help to convert the raw biomass into a solid product that has improved storage, transport, and handling characteristics. The commonly used densification technologies that are used for biomass are pelleting, briquetting, cubing, extrusion, and agglomeration. Pelleting and briquetting have been successfully used for many years in many countries. All produce densified products with different quality attributes. The quality attributes of the densified product depend on the biomass, physical properties, chemical composition, and densification process variables. The focus of this chapter will be understanding how processing these feedstocks using different densification systems and process variables impact their overall quality attributes, such as unit, bulk, and tapped density, and durability.

Keywords Densification systems · Particle binding mechanism · Process variable · Densified product quality attributes

2.1 Introduction

Biomass, as harvested, has less bulk density, which makes storage, feeding, and transportation difficult. Typical bulk densities of herbaceous biomass after harvesting and baling are about 150–160 kilograms per cubic meter (kg/m^3). It can be further reduced in size by using a grinder, typically around 80–150 kg/m^3, based on the screen size used in the grinder. The woody biomass after harvesting and chipping is about 200–220 kg/m^3 (Tumuluru et al. 2016). Two major densification limitations in using biomass before it is converted include its extreme difficulty to move from one place to another during feeding, which results in bridging of the biomass particle, and storage, which requires a large amount of space due to its low density. Another major limitation with biomass is its variable moisture content, which results in dry matter losses during storage that greatly influences grinding energy, throughput, and particle size distribution. Typically, high-moisture content greater than 20% (w.b.) results in a need for higher grinding energy, which in turn

© Springer Nature Switzerland AG 2020

J. S. Tumuluru, *Biomass Densification*,
https://doi.org/10.1007/978-3-030-62888-8_2

leads to irregular particle sizes and shapes, increased grinder throughput, and non-uniform particle size distribution (Tumuluru and Yancey 2018), thereby directly causing inconsistent mass and heat transfer in conversion. According to Tumuluru (2018a), variability in the moisture content of the biomass results in preprocessing and conversion issues. The particle size variability (coarse, medium, and fine particles in the grind) causes plugging in storage bins and augurs, not completely cooking effectively in the digesters, and the subsequent plugging of the downstream equipment—or particles that are too small, which can cause fires and explosions, the plugging of the weep holes of the feeder, inconsistent heat and mass transfer during conversion, and detrimental influence on the ash content and composition, as well as serious health hazards for workers.

2.1.1 Densification

One way to address the variability issues related to moisture, particle size, and shape in the biomass can be by mechanical densification. During mechanical densification, the biomass is compressed and extruded, and the particles are densely packed in the die, which reduces the void spaces between the particles. Densely packing the particles in the die increases the bulk density of the biomass by about four to five times (Tumuluru et al. 2011). The conventional mechanical densification systems that are used in biomass processing include the pellet mill, the briquette press, and the cubers (Tumuluru et al. 2011). These same authors state that densification converts biomass into an aerobically stable feedstock with consistent size, shape, and increased density. In addition, they reasoned that biomass densification is critical to improving handling and transportation efficiency to biorefineries. It also produces a product with consistent size, shape, density, and moisture content. In fact, the components in the biomass during the densification process get fractionated, which further improves the performance of the feedstock during biochemical and thermochemical conversions. Finally, the densified material can be designed for predetermined specifications desired for conversion and logistics systems.

2.1.2 Bonding Mechanism During Densification

Tabil (1996) and Tabil and Sokhansanj (1996a, b) suggested that the elastic and plastic deformation of the particles during the application of pressure results in particle binding. This is one of the most critical aspects of the pelleting to form pellets with the desired density. Many authors have concluded that the agglomeration of the particles during densification is due to the formation of solid bridges because of the chemical reactions between the biomass components (mainly the binders), binder hardening, and the solidification of melted substances (Rumpf 1962; Sastry and Fuerstenau 1973; York and Pilpel 1972; Pietsch 1984). Most of the

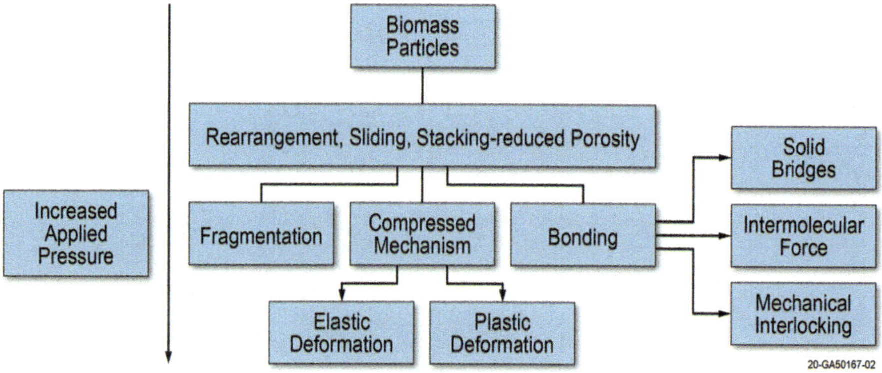

Fig. 2.1 Deformation mechanisms of powder particles under compression (derived from Comoglu 2007, and Denny 2002)

reactions that happen in biomass are primarily due to the presence of moisture in the biomass. Moisture, in the presence of temperature and pressure, results in modification of the chemical components in the biomass, thus resulting in particle bonding. According to Sastry and Fuerstenau (1973), Pietsch (1984), and Ghebre-Sellassie (1989), moisture in the pendular state fills the voids. In the funicular state, the liquid content increases and reduces the pore volume. In the capillary and droplet state, the agglomerate is completely covered by the liquid, where the surface tension is the primary force to hold the primary particles. According to Schineberger (1971) and Sherrington and Oliver (1981), the particle attraction is due to van der Waal's electrostatic or magnetic forces and inter-particle attraction, which contribute to particle bonding. As proposed by Comoglu (2007) and Denny (2002), Fig. 2.1 suggests there are three stages in the powder compaction process. In the first stage, the particles rearrange and retain most of their properties, while in the second stage, these particles are forced against each other and undergo plastic and elastic deformation. Also at this stage, the particle rheological properties and morphological characteristics are influenced where particle bonding occurs due to van der Waal's electrostatic forces. In the third and final stage, reduction volume occurs due to the application of higher pressure, where the density of the pellet reaches the true density of the individual components of the powders. Also, at this stage, the particles that have undergone deformation cannot change their position in the compact due to reduced activities. The densification or compaction process is governed by the process variables and powder properties. Controlling the process variable, which influences the inter-particle cavities, helps to produce a desirable product.

The composition is another variable that has a significant impact on the binding behavior of biomass. Chemical compositions—such as cellulose, hemicellulose, protein, starch, lignin, crude fiber, fat, and ash—impact the type of chemical bonds that are formed during densification. According to Briggs et al. (1999), the protein and starch plasticize at higher compression forces and temperatures, which act as a binder to form a strong pellet. Wood (1987) and Thomas et al. (1998)

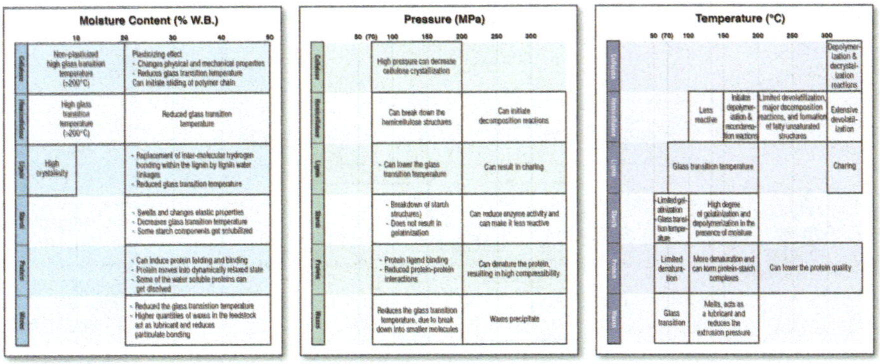

Fig. 2.2 Changes in biomass components due to the application of moisture, temperature, and pressure (derived from Tumuluru 2018a)

concluded that the starch in the biomass gelatinizes and results in better binding. In addition, the lignin in the lignocellulosic biomass softens in the presence of moisture, temperature, and pressure, which acts as a binder. Tumuluru (2018a) developed a figure to understand how moisture, temperature, and pressure affect various biomass contents—such as cellulose, lignin, starch, portion, and waxes, as observed in Fig. 2.2. The figure clearly indicates that various biomass components have different impacts when subjected to pressure, temperature, and moisture. Studies conducted in understanding the glass transition (Tg), temperature of biomass components using dynamic mechanical analysis (DMA), and solid-state nuclear magnetic resonance (SS-NMR) have indicated that the Tg of lignin and starch is greatly influenced by temperature and moisture content (Tumuluru 2018a), as observed in Figs. 2.3 and 2.4. The DMA curve indicated that a low preheating temperature of about 50–70 °C could help to activate some of the biomass components, such as lignin, waxes, water-soluble carbohydrates, protein, and others to take an active part in the densification process. The solid-state NMR spectrum has indicated that the Dipolar dephased spectrum of wet biomass shows xylans and lignin as the mobile components, while the ^{13}C spin-lattice relaxation rate constants show that the hemicellulose mobility is increased more than the lignin in the presence of water, as observed in Table 2.1.

Compared to ceramic and pharmaceutical powders, biomass has natural binders, such as lignin, protein, and starch. In addition, the heterogeneous nature of biomass further complicates our understanding of the binding phenomena. According to Bilanski and Graham (1984) and O'Dogherty and Wheeler (1984), the higher pressure applied during densification can crush the biomass and open up the cell structure, which makes some of the components in the biomass—such as protein and pectin—act as binders. Several authors (Tabil and Sokhansanj 1996a, b; Adapa et al.

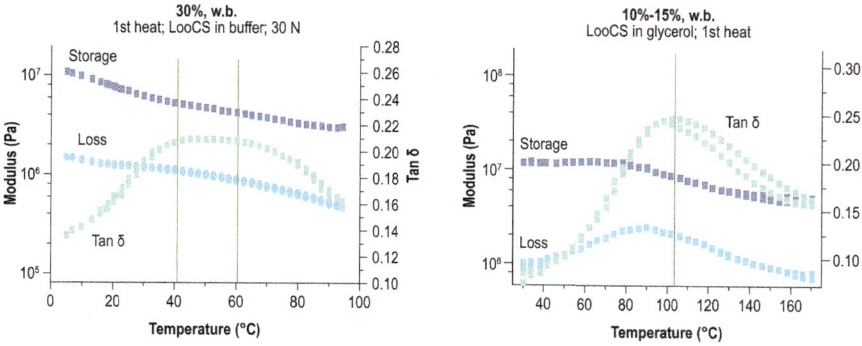

Fig. 2.3 Compression-torsion DMA of ground corn stover immersed in an aqueous buffer with a pH of 7 (left) and in glycerol (right). Thermal transitions are defined by peaks in the tan δ curve (tan δ = loss modulus ÷ storage modulus) (derived from Tumuluru et al. 2017)

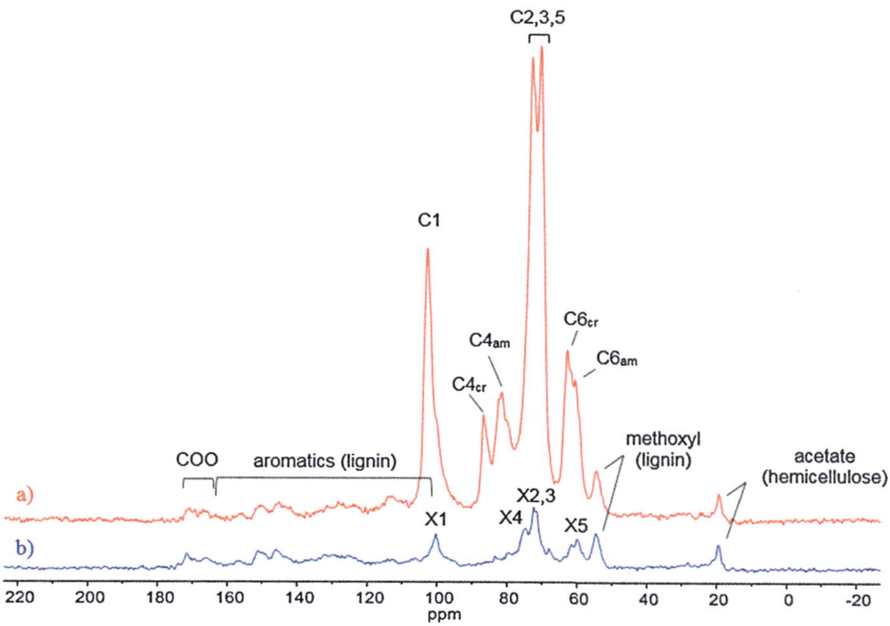

Fig. 2.4 (**a**) Solid-state NMR spectrum of wet corn stover and (**b**) spectrum with dipolar dephasing (derived from Tumuluru et al. 2017)

2002, 2009; Mani et al. 2004) concluded that starch, lignin, and protein takes an active part in the pelleting of alfalfa, wheat, and barley grind. Kaliyan and Morey (2006) used scanning electron microscopy (SEM) to understand the solid-type bridges formed due to pelleting and briquetting of corn stover and switchgrass.

Table 2.1 ^{13}C T_1 mobile component relaxation rate constants

	Chemical shift (ppm)				
	72.7 (HC)	70.4 (HC)	62.8 (HC)	53.6 (LI)	19 (HC)
Dry corn stover	12 s	12 s	4.2 s	3.0 s	10 s
Wet corn stover	4.5 s	4.6 s	1.6 s	2.3 s	6.5 s

Derived from Tumuluru et al. (2017)

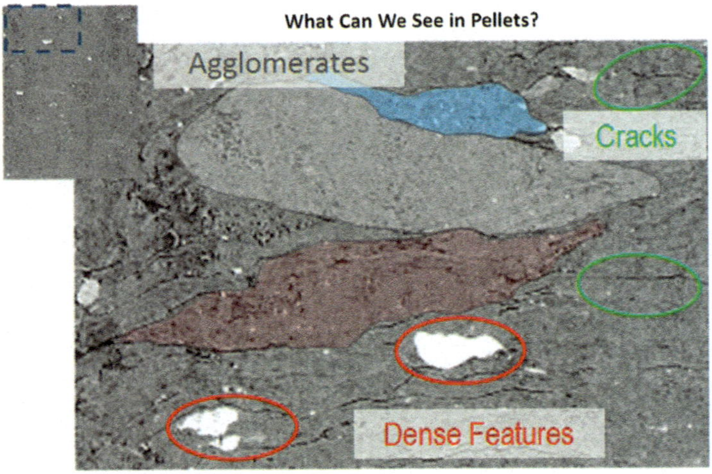

Fig. 2.5 CT scan images of a corn stover pellet (derived from Tumuluru et al. 2019, 2020)

Recent studies by Tumuluru et al. (2019) using computed tomography (CT) scan and focused ion beam (FIB) tomography has helped to define the agglomeration behavior of corn stover particles, in addition to microstructure formation due to the impact of process variables, such as moisture content, grind size, compressive force, and residence time. Advanced characterization of the pellets using an X-ray CT scan, which is a 3D non-destructive imaging process that can enable spatial and morphological characterization without destroying the pellet and can help to understand the particle agglomeration behavior (Tumuluru et al. 2019). Some of the CT scans and SEM images that help us to understand the particle agglomeration and lignin mobilization during pelleting are given in Figs. 2.5 and 2.6. It is very clear from the CT scan image that after densification, the particles come closer and form a dense compact with a clear indication of quarks and void spaces. In addition, it is possible to see the dense features in the pelleted corn stover (see Fig. 2.5). The SEM image indicates that during the pelleting process, the biomass structure undergoes significant changes (see Fig. 2.6). The cellular structure is largely disrupted but still shows residual evidence of the previously existing structure. In addition, the SEM images indicate that stomata and villi are still visible after pelleting. Stomata guard cells tend to have high-lignin content, which remains intact after the pelleting process, indicating only a partial mobilization of all the lignin present during the pelleting

Fig. 2.6 SEM images of pelleted corn stover (derived from Tumuluru et al. 2019, 2020)

process. More studies at a micro-level using these techniques will be useful in understanding the interaction of feedstock and process variables on the microstructure formation in the pellets.

2.1.3 Densification Systems

The densification systems currently used by the biomass industry are baling, pelleting, extrusion, briquetting, and agglomeration. Among the various densification methods used by the industry, pelleting and briquetting are the most common. These include the high-pressure densification process where biomass particles are bound together without any binder. These technologies are also referred to as "binderless technologies" (Sokhansanj et al. 2005).

2.1.3.1 Pellet Mill

Pellet mill has perforated hard steel die with two or three rollers. Rotating die or rollers force the feedstock through the perforations to form densified pellets. In pellet mills, the raw biomass is compressed and extruded to produce smaller densified products called pellets. Fig. 2.7 shows the various pellet mill components used for the continuous production of pellets (Tumuluru et al. 2011). Two types of pellet mills exist: the flat die pellet mill and the ring die pellet mill. The principle of operation in both are the same where the raw biomass is compressed and extruded, but the quality of the pellets, die dimensions, and how the raw material is fed into the die is different.

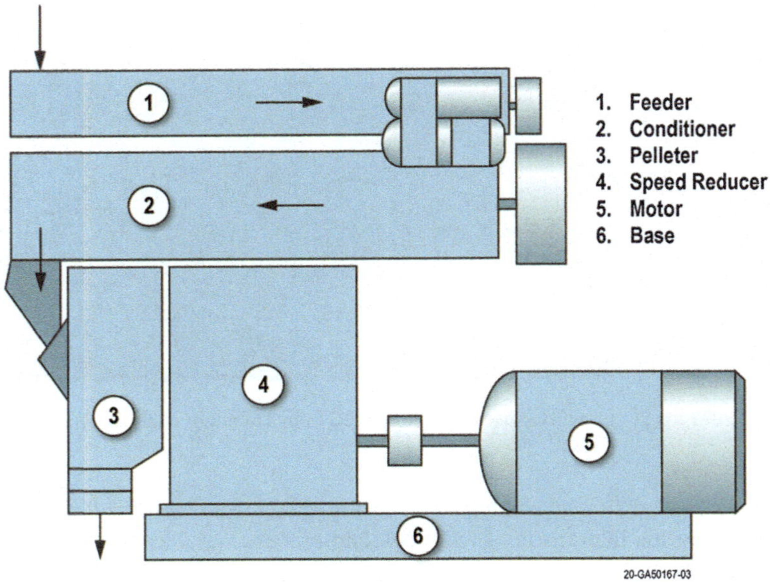

Fig. 2.7 Pellet mill components (derived from Tumuluru et al. 2011)

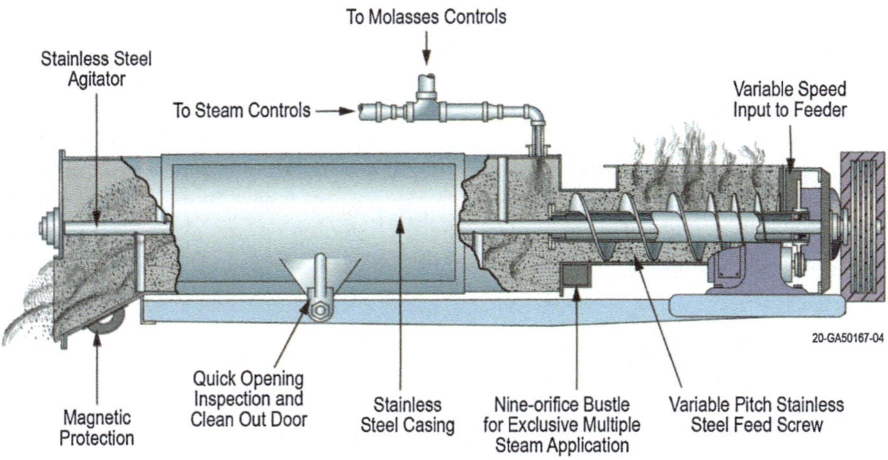

Fig. 2.8 Various components of the screw conditioner (derived from Tumuluru et al. 2010a)

2.1.3.2 Preconditioning of Biomass

Biomass is initially fed into screw conditioners where the biomass is pretreated with steam. Figure 2.8 indicates the various components of the screen conditioner. The addition of steam helps to soften some of the biomass components, such as mobilizing the lignin and gelatinizing the starch, which results in better binding of the particles in the pellet die. The steam-conditioned biomass is further discharged into the feed spout, which leads to the pellet die. The flights that are provided in the die cover help to feed the biomass evenly in each of the rollers, and the feed distributors spread the feed evenly on the face of the die. Some pellet mills are provided with force-feed augurs, which can help to move the material into the die more uniformly and consistently. In addition, force-feed augurs help to feed biomass material that is low and variable in density, such as municipal solid waste (MSW). Pellet mills are further provided with knives to cut the extruded pellets to their desired lengths.

2.1.3.3 Flat Die Pellet Mills

Flat die pellet mills come in two types—rotating rollers and rotating dies. In the rotating roller, the roller is moving where the die is stationary, whereas, in the rotating die, the pellet die moves where the rollers sit on the die. Two more rollers are typically used in flat die mills, where the biomass is dropped by its own weight into the pellet mill chamber where it gets compressed between the rollers and the die and extrudes pellets through the die hole. Figure 2.9 indicates the flat die pellet mill and the extrusion of pellets through the die. The major advantages of the flat die pellet mill are its small size and light weight, its compact design, its relatively easy cleaning and maintenance capacity, and its ability to be used for both feeding and producing pellet fuels. The major challenge with flat die mills is adjusting the gap

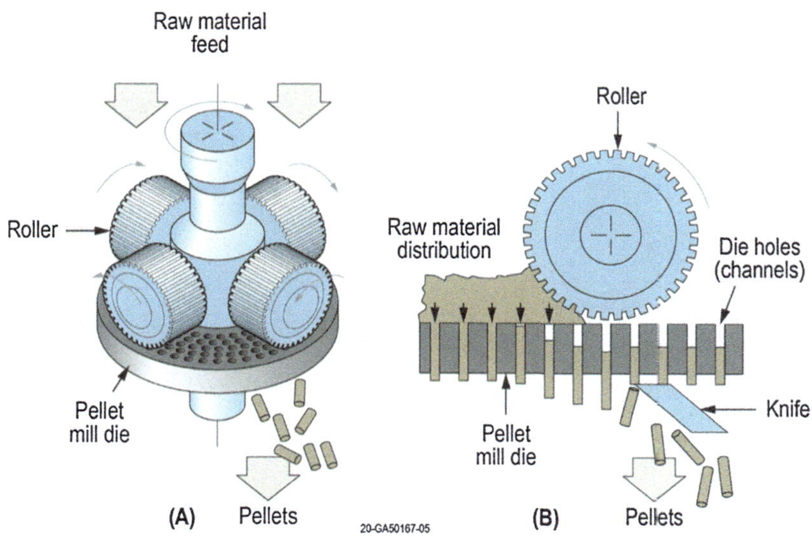

Fig. 2.9 Design and operating principles of the flat die pellet mill (derived from Gemco Energy 2019)

between the rollers. In addition, the die can be tricky and needs to be adjusted based on the type of raw materials being fed into it.

2.1.3.4 Ring Die Pellet Mills

The basic principle of pelleting biomass in a ring die pellet mill is its distribution over the inner surface of the rotating perforated die ahead of the rollers, which is further compressed into the die holes to form the pellets. Figure 2.10 indicates the pelleting mechanism in a ring die pellet mill. In general, a ring die pellet mill is used for the production of fuel and feed pellets. Ring die pellet mill generate less wear and tear as the roller travels the same distance and are more energy-efficient than flat die pellet mills that are available in higher capacities. The major drawback with using a ring die pellet mill is the slip of the biomass during squeezing into the pellet die. Proper die design can help to overcome this issue. Table 2.2 indicates the comparison of flat die and ring die pellet mills.

2.1.3.5 Briquette Press

There are two types of briquette presses: the hydraulic press and the mechanical press. Mechanical presses generate more pressure than hydraulic presses. Mechanical piston presses are used for large-scale production of densified biomass. In the briquetting process, the plunge connected to the continuously rotating center mass

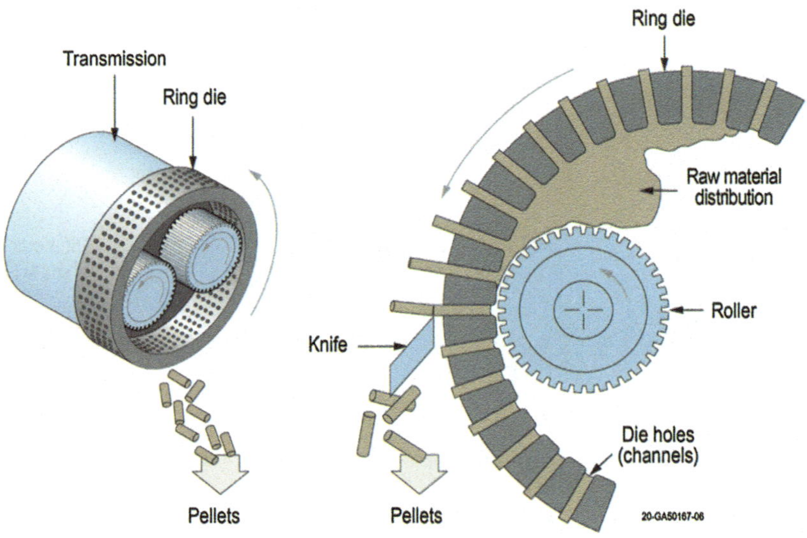

Fig. 2.10 Design and operating principle of the ring die pellet mill (derived from Gemco Energy 2019)

Table 2.2 Comparisons between the flat die and ring die pellet mills

	Flat die pellet mill	Ring die pellet mill
Cost ($)	Low	High
Throughput (ton/hr.)	Low capacity mills	High-capacity mills
Feeding mechanism	By gravity and material weight	Forced feed
Pressure	Adjustable because the diameter of the rollers is not limited by the diameter of the pellet die	Limited pressure because the diameter of the rollers is limited by the diameter of the pellet die
Damage	Less damage due to low rotational speed	More damage due to high-rotary speed of the ring die
Wear	More	Less
Energy efficiency	Less energy efficient	More energy efficient
Slide slip of material	Lower	Higher

Derived from Gemco Energy (2019)

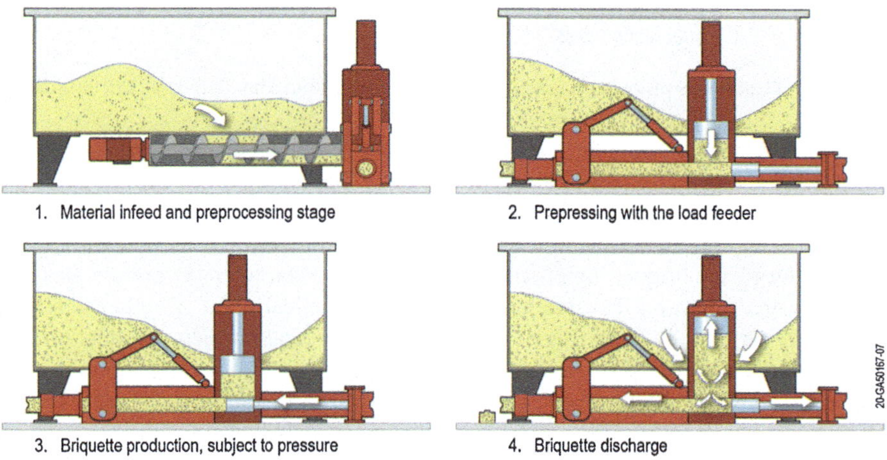

Fig. 2.11 Briquetting of biomass in a C-150 briquette press

forces the biomass through a restricted opening. Electric or hydraulic motors are typically used. For commercial-scale production, electric motors are used, whereas, for small- and medium-scale production, hydraulic motors are used. The challenge with hydraulic motors, however, is when the oil heats up. Some hydraulic presses are provided with cooling fans to reduce the heat in the hydraulic oil. A typical working mechanism in a hydraulic briquette press is given in Fig. 2.11. Additionally, the four different stages needed for processing biomass in a hydraulic briquette press are also shown in Fig. 2.11. Initially, the material is preprocessed in the augur connected to a hopper. In the next stage, the preprocessed biomass is pressed with a load pusher. In

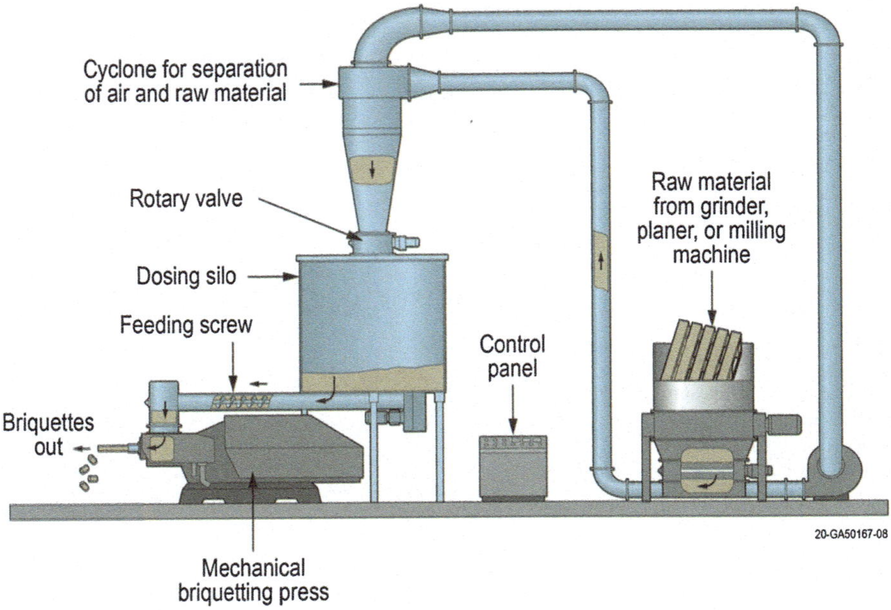

Fig. 2.12 Continuous piston-type briquetting press (derived from Tumuluru et al. 2010a)

the third stage, the briquette is produced via the application of pressure using a hydraulic ram. Finally, the formed briquette is discharged from the press in the fourth stage through the die opening.

In the case of a mechanical press, which is driven by electric motors, the loss of energy during the briquetting process is less than that seen with a hydraulic press. In addition, the return on investment by using a mechanical press is higher than that seen with a hydraulic press. Figure 2.12 shows the typical briquette production in a mechanical press. In general, for the briquetting process, larger particle size is preferred as the binding of the particle is more due to mechanical interlocking. The major application of produced briquettes is used for heating furnaces, which tend to be much better than those observed with conventional log use because they have higher heating value, less particulate emissions, and they can be stored and transported more efficiently due to a more uniform size and shape, in addition to a higher density.

2.1.3.6 Screw Extruder

In a screw extruder, a screw rod is used to extrude biomass material where the extruder barrel is heated from outside to maintain a constant temperature during compression and extrusion. During extrusion, the biomass undergoes mixing, shearing, and compression. During extrusion, the lignin and starch content in the biomass

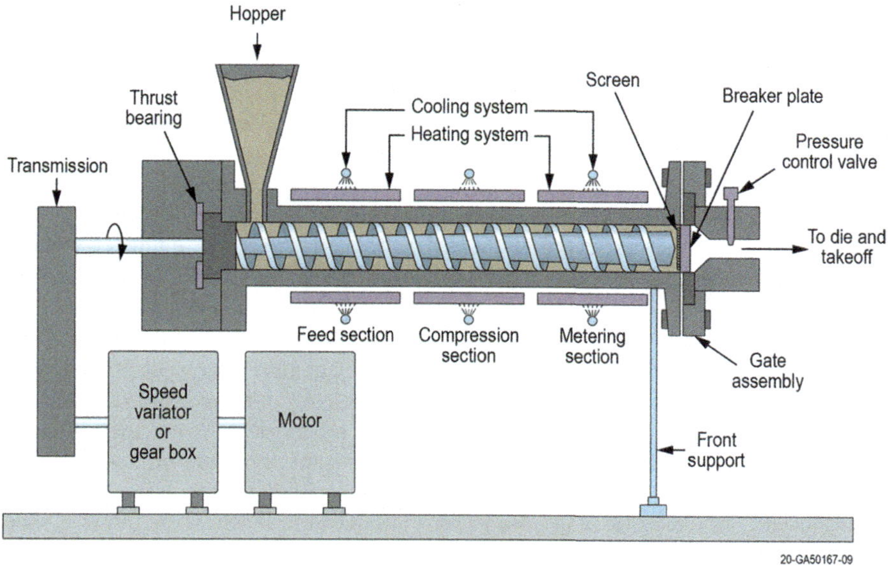

Fig. 2.13 Typical extruder for pelleting applications

is softened, which helps to form a densified biomass. There are two types of extruders: single- and twin-screw extruders. Single- and twin-screw extruders are typically used in the food, feed, and polymer industries. The twin-screen extruder has two mutual clenching shifting screw rods, which are mainly used to bring significant changes to the physical properties of the feedstock, in terms of size, shape, surface area, morphological characteristics, and chemical composition. During the extrusion process, when the biomass material is passed through the extruder, the shearing and mixing occur due to the wall and internal friction in the biomass. This results in an increase in internal temperature in the extruder. If the generated temperature is not sufficient, then external heaters are provided to maintain the extruder barrel at the desired temperature. The heated, sheared, and mixed biomass is further extruded through a constricted die hole to form densified biomass; as such, the design of the screw has a great impact on the resulting quality of the extruded products. The most common screw design configurations are the helix angle, length to diameter (L/D) ratio (also known as the compression ratio), and a number of mixing elements. Figure 2.13 shows a typical single-screw extruder used in the food, feed, and biomass processing industries (Crowley et al. 2007). Extruders are divided into different types, such as high- and low-shear extruders. High-shear extruders are used to produce precooked, gelatinized, and heat-treated products. These extruders are classified as high-temperature, short-time extruders, where the biomass is preheated with steam or hot water and processed further (Harper 1981). Low-shear extruders are equipped with moderate-shear, high-compressed, and grooved barrels to enhance mixing. These extruders are used for low viscous materials (Harper 1979, 1981).

2.1.3.7 Roller Press

The primary operating principles for a typical roller press are pressure and agglomeration. The press is provided with two rollers of the same diameter rotating horizontally in opposite directions, as observed in Fig. 2.14. These two rollers are arranged in such a way that there is a small gap between them. When the raw material is passed through the gap of the rollers, a briquette is formed. Various process variables impact the quality of the produced briquette, such as the roller diameters, the gap between the force of the rollers, and the shape of the die. According to Yehia (2007), the gap between the rollers depends on many factors, such as the type of the biomass, the particle size, the moisture content, and the addition of binders. The shape of the densified biomass depends on the type of die being used. For example, a smooth roller is used if the machine output is a sheet having a specific thickness. Currently, researchers are looking at developing new roller presses with multiple rollers that can produce high-quality densified products, as seen in Fig. 2.15. These new designs are aimed toward pre-agglomerating the biomass before it is fed to the roller press. Studies conducted by Larsson et al. (2008) indicate that pre-compaction before pelleting helps to enhance the stable reed canary grass pellet production process.

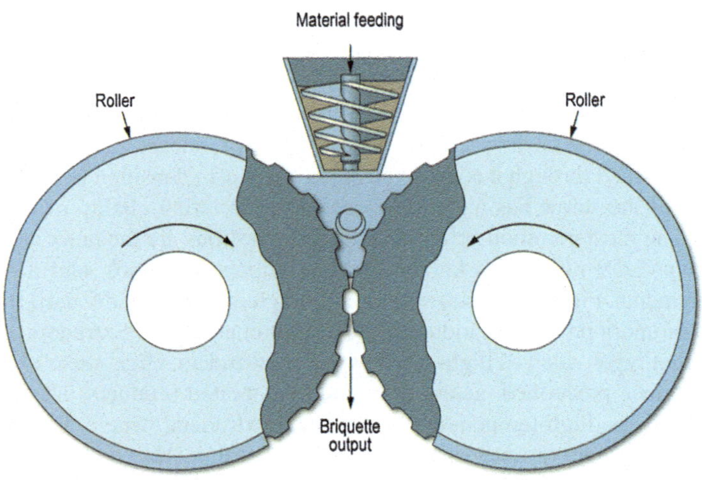

Fig. 2.14 Roller press for briquette production

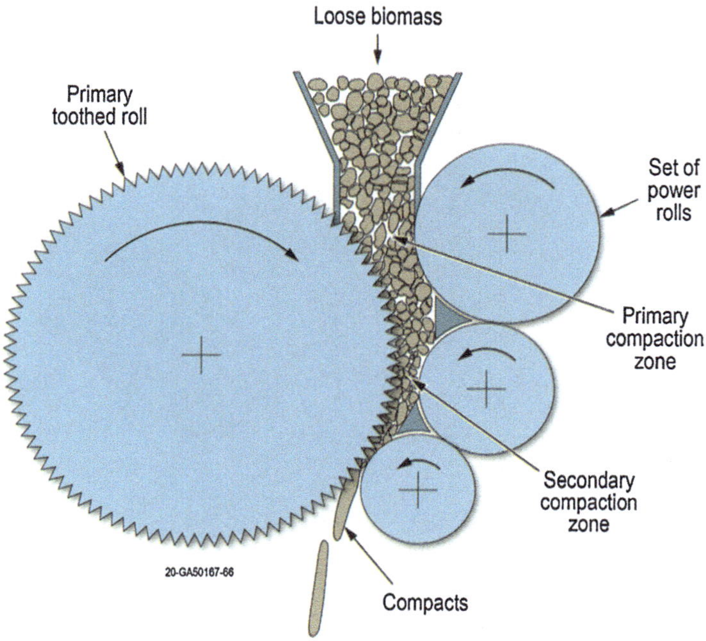

Fig. 2.15 New conceptual roller press design for biomass densification (derived from Sokhansanj et al. 2005)

2.1.3.8 Agglomerator

The term "agglomeration" refers to the particle size growth of the powders by using either natural or chemical binders. Agglomeration technologies are widely used in the pharmaceutical, chemical, and mineral industries. For example, hydrated lime, pulverized coal, iron ores fly ash, cement, and other minerals are commonly agglomerated using binders. In the ore industry, disc granulators or pelletizing discs are commonly used. These systems work on a tumbling technique principle of agglomeration. In this process, inclined discs are quite commonly used. The resulting feed is usually a fine powder mixed with a liquid—most commonly water. The particles collide with each other in a spinning disc. If the bonding forces are stronger than the repulsion forces, the particles coalesce and grow. In this process, the agglomerate grows either by coalescence or by layering. The major forces that result in this binding are van der Waals, magnetic, and electrostatic. The agglomeration is also strongly dependent on the weight of the solid. If the weight of the solid is higher, it can work against agglomeration. The various process variables that impact the agglomeration of powders are: (a) pan-tilt angle; (b) rotation speed; (c) powder-to-binder feed ratio; (d) powder feed rate; (e) feeder points; (f) rim height; and (g) scraper position (Tumuluru et al. 2010a). According to Mani (2014), agglomerates are produced due to: (a) the attraction of forces between

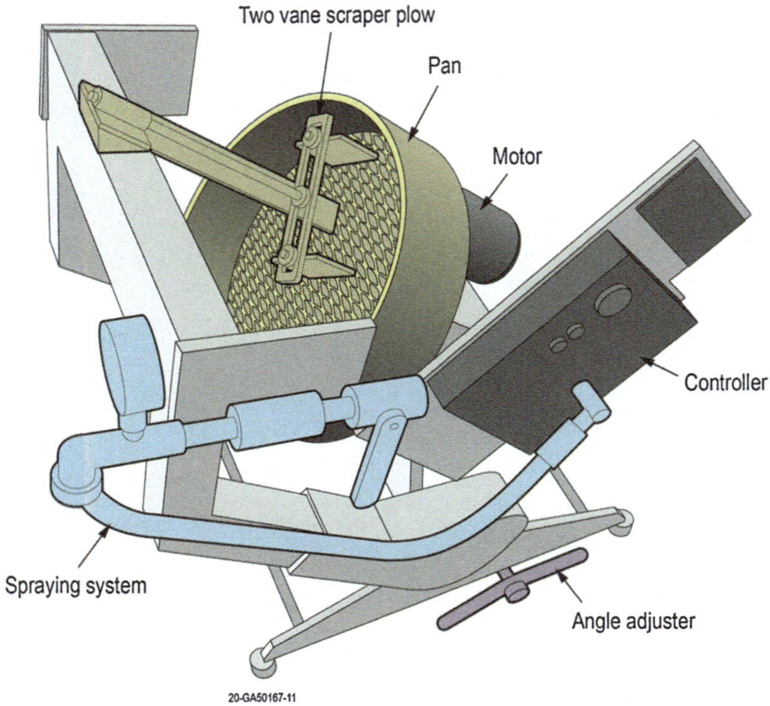

Fig. 2.16 Pan type agglomerator

solid particles; (b) interfacial forces and capillary pressure in movable liquid surfaces; (c) adhesion and cohesion forces in non-freely movable binder bridges; and (d) mechanical interlocking. Studies on the agglomeration of agricultural or woody biomass are scarce. The process of biomass agglomeration consists of the following steps: (a) fine grinding of biomass; (b) mixing the finely ground biomass with a binder; (c) agglomeration of finely ground biomass with a binder; and (d) drying of the agglomerates. Figures 2.16 and 2.17 show the pan- and disk-type agglomerator typically used by the industry. Studies conducted by Mani (2014) on the agglomeration of lime-treated switchgrass and torrefied wood powder using a corn starch binder at different concentrations (e.g., 2.5, 5.0, 7.5 wt. % [d.b.]) indicated that lime-treated granules have a bulk density of 175–194 kg/m^3, while torrefied wood powder granules showed a bulk density of 330–387 kg/m^3. Figure 2.18 shows an example of granules produced at different conditions using torrefied wood powder.

2.1.3.9 Tablet Press

A tablet press is provided with a hydraulic motor, which turns and feeds the 4–6-inch diameter cylindrical mold. Figure 2.19 shows a single punch tablet press (Kumar

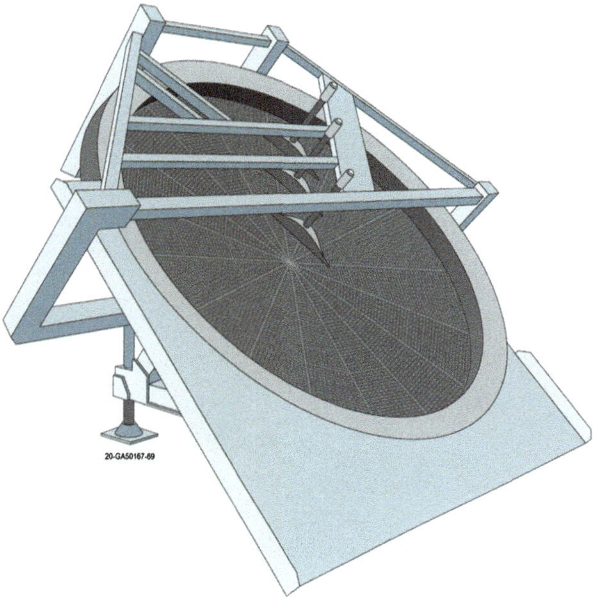

Fig. 2.17 Disk agglomerator typically used in the industry

Fig. 2.18 Granulation of torrefied pine chips with a starch binder using (**a**) raw pinewood starch granules; and torrefied pinewood starch granules at (**b**) 250 °C; (**c**) 275 °C; and (**d**) 300 °C (derived from Mani 2014)

2011). The process of a tablet press includes three stages: (a) filling; (b) compression; and (c) ejection. A single tablet press is also called an eccentric press or a single station press. In this press, the compaction force on the material is exerted by the upper punch only. Dies define the size and shape of the table formed. The major

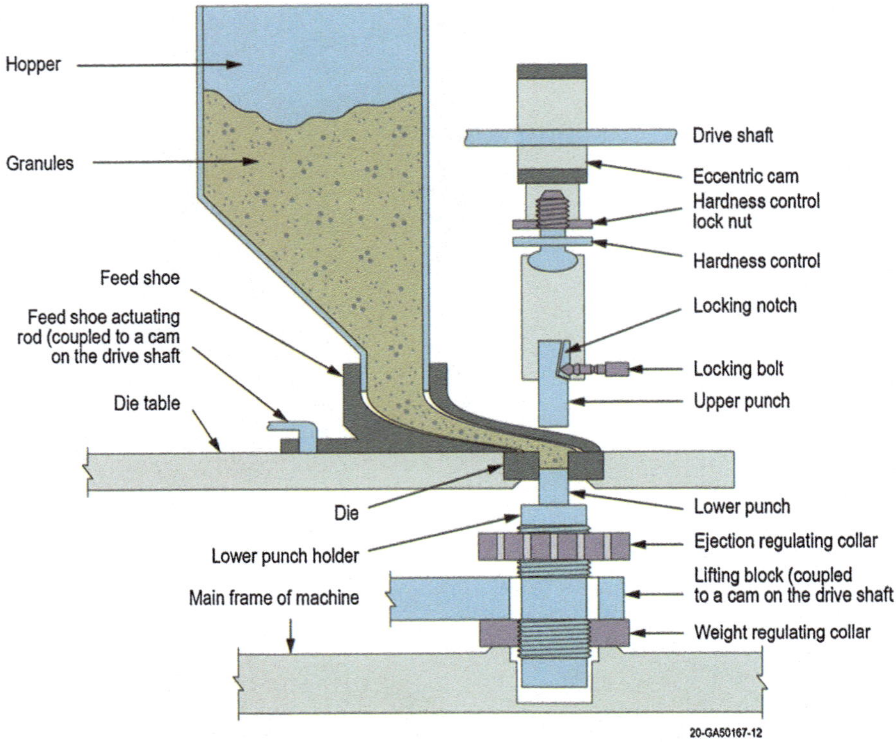

Fig. 2.19 Energy tablet-making machine from biomass

advantages of this press are as follows: (a) a small structure; (b) a high utilization ratio; (c) high-pressure use capability; and (d) raw material loading control. Typically, very high pressure of 20,000 pounds per square inch (psi) is applied during compression, which results in the binding of particles without having to add binders. The application of using a tablet press for biomass was completed by the Bradford Research and Extension Center at the University of Missouri. Figure 2.20 shows a sample tablet produced using corn stover. In this process, a larger particle can be used in the binding due to mechanical interlocking and lignin mobilization because of the high pressure. The application of high pressure in the range of 20,000 psi results in high-unit density tablets (55 lb./ft^3). Currently, tests are being conducted to understand how these biomass tablets perform in power plant and gasification applications (Gibson 2010).

2.1.3.10 Energy Consumption of the Densification Systems

Tumuluru et al. (2010a) and Tumuluru et al. (2011) discussed the energy consumption of various densification systems. The results indicated that feedstock type and

Fig. 2.20 Densified biomass produced using a tablet press. (Source: Bradford Research and Extension Center, University of Missouri)

the specific equipment used for densification both influenced the energy consumption of the process. According to Mewes (1959), 37–40% of the total energy needed for doing this is required to compress, where the remaining energy is needed to overcome frictional resistances. Studies conducted by Tumuluru et al. (2011) indicated that for moist biomass (around 15%), energy consumption decreased as compared to that needed for lower moisture materials (<10%, w.b.). In addition, at very low-moisture content, a lot of steam must be added to the biomass before it goes into the pellet mill. Studies conducted by Aqa and Bhattacharya (1992) indicated that densification process variables, such as die temperature and preheating temperature of the sawdust, have an impact on the quality and energy consumption of the process. Their research focused on the quality and energy consumption of the briquettes produced when sawdust is preheated to 115 °C and concluded that the energy inputs to the briquetting machine motor, die heaters, and overall system reduced by 54, 30.6, and 40.2%, respectively. These authors also concluded that higher throughputs through the briquetting machine reduced the electrical energy requirements per kilogram (kg) of sawdust. Additional studies conducted by Tumuluru et al. (2017) on pelleting corn stover at moisture content values in the range of 13–30% (w.b.) indicated there is a slight increase in energy consumption up to 22% (w.b.) moisture content, whereas further increasing the moisture content values also increased energy consumption exponentially, as observed in Fig. 2.21. Tumuluru (2015) discussed how pelleting energy changes in a flat die pellet mill for corn stover pelleted at 28–38% (w.b.) moisture content values, as shown in Fig. 2.22. They concluded that higher rotation speed increases the throughput of the mill, thereby decreasing the specific energy consumption of the process. Also, they reported that increasing the preheating temperature reduces the specific energy consumption of the pelleting process. Tumuluru et al. (2016) studied the effect of a starch-based binder impact on the energy consumption of corn stover pelleting

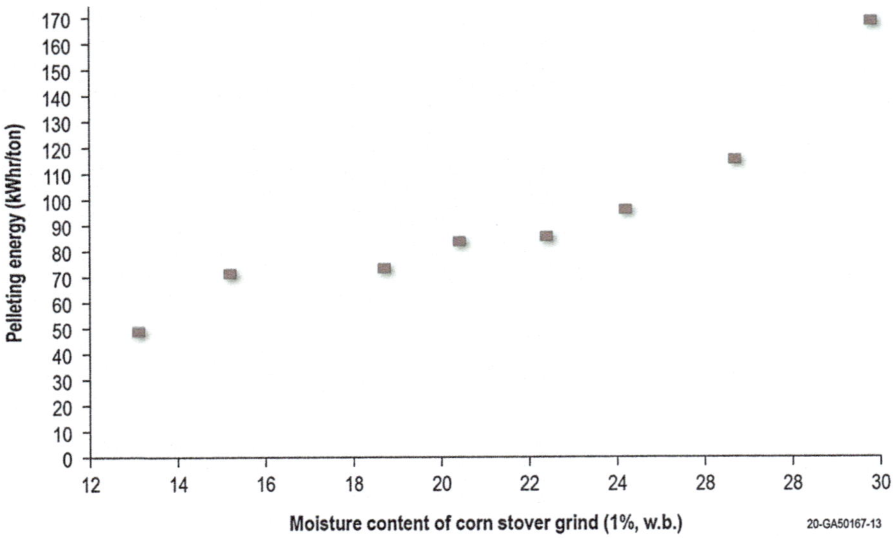

Fig. 2.21 Energy consumption of the pelleting process with respect to corn stover moisture content in a ring die pellet mill with a 1 ton/hr. throughput

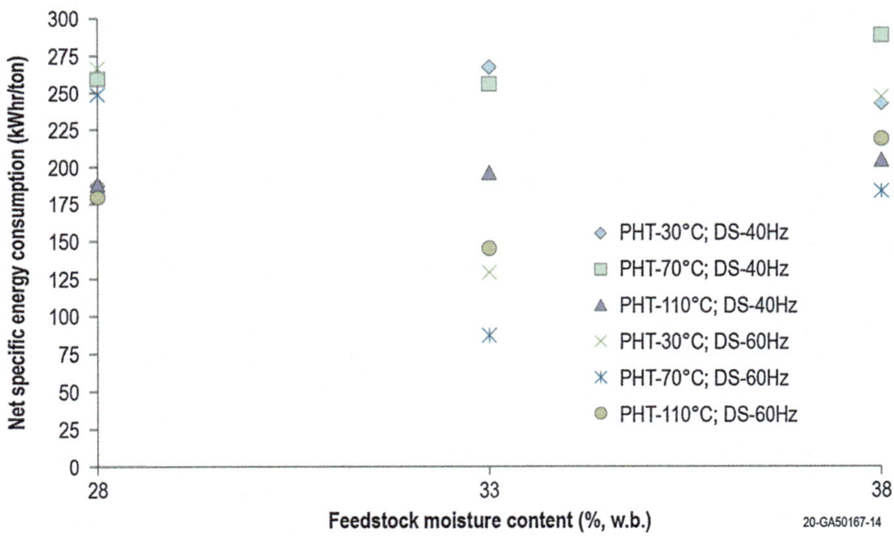

Fig. 2.22 Effect of preheating, die speed, and feedstock moisture content on net specific energy consumption (derived from Tumuluru 2015)

process in a flat die pellet mill (Fig. 2.23). Their findings indicated that adding 4% starch almost reduced the energy consumption by half. Similar results were observed by Tumuluru et al. (2016) on the pelleting of woody biomass at higher moisture

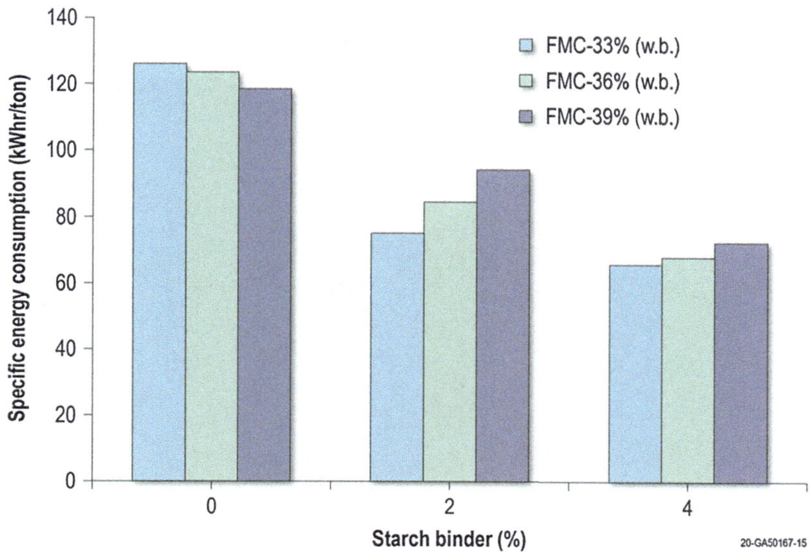

Fig. 2.23 Effect of feedstock moisture content and starch binder on specific energy consumption

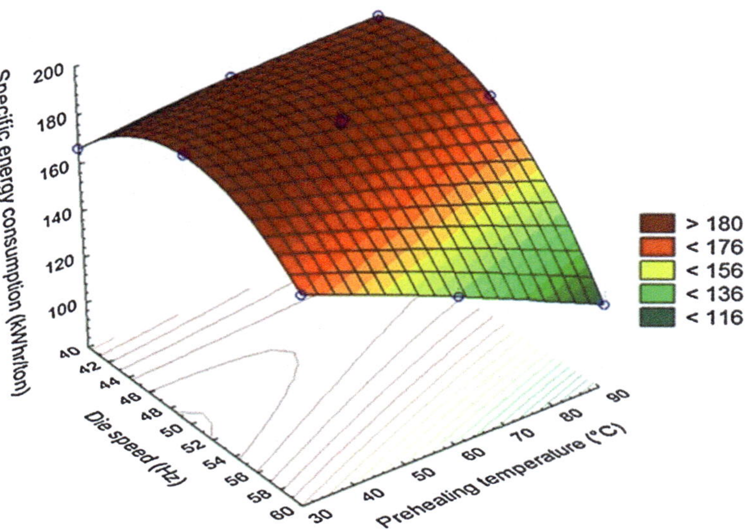

Fig. 2.24 Effect of die speed and preheating temperature on specific energy consumption

content values of 33–39% (w.b.), where higher preheating temperatures and higher die speeds reduced the specific energy consumption of the process, as observed in Fig. 2.24. In the case of the briquette press, Tumuluru (2019a) has suggested that moisture content and the particle size of the grind has a significant impact on the

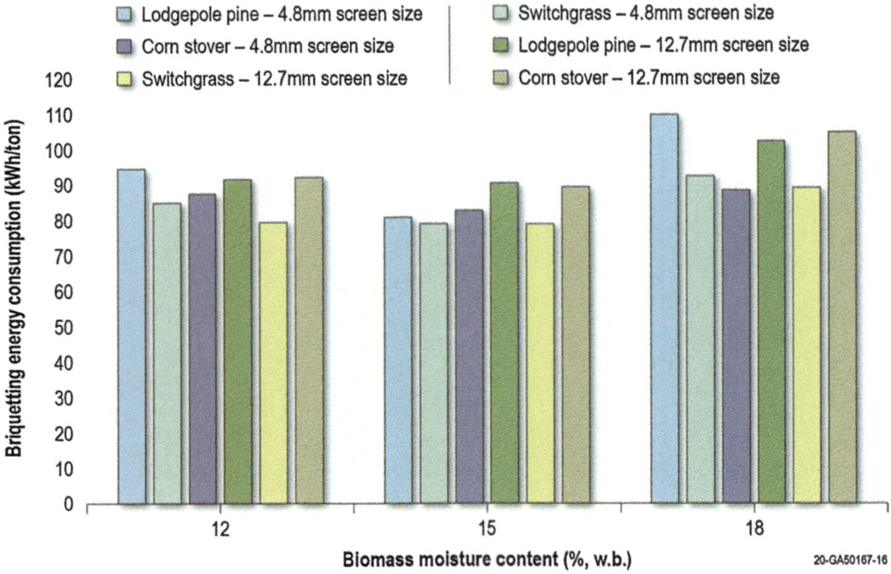

Fig. 2.25 Energy consumption of the briquette production process at different process conditions

energy consumption of the process, as observed in Fig. 2.25. Their results indicated that with a larger particle grind size of 12.7 mm and a higher moisture content value of 18% (w.b), the energy consumption of the process increased to around 110 kilowatt-hours per ton (kWh/t).

Tumuluru et al. (2011) discussed the suitability of densified products for various end-use applications. They also reported the optimum process conditions suitable for different densification systems. This data was updated in a revised table based on the conclusions obtained from these researchers, with additional data published in the literature, as observed in Table 2.3. Tumuluru et al. (2011) also concluded that using a pellet mill and a screw press help to produce a homogenous product. The major reason for this is that by using both processes, the densification of raw material gets mixed into the die before being extruded, and because the extruder can go to high temperatures, carbonization of the resulting extrudate is possible. Most of the final products can then be used for both biochemical and thermochemical conversion applications in the production of biofuels. Their review also indicated that wear is higher for the pellet mill, screw extruder, and roller press as compared to a typical piston and agglomerator. Table 2.4 provides updated information regarding the comparison of different densification systems for biofuels production.

Table 2.3 Specific energy consumption for the densification of different feedstocks

Materials	Type of densification unit	Specific energy consumption (kWh/t)
Sawdust	Pellet mill	36.8
Municipal solid waste	Pellet mill	16.4
Bark + wood	Pellet mill	30–45
Straws + binders	Pellet mill	37–64
Straws	Pellet mill	22–55
Grass	Pellet mill	33–61
Switchgrass	Pellet mill	74.5
Alfalfa	Pellet mill	30
Straws + binders	Cubing machine	75
Grass	Cubing machine	28–36
Cotton trash	Cubing machine	60
Hay	Cubing machine	37
Sawdust	Piston press	37.4
Straws	Screw press	150–220
Grass	Piston press	77
Straws + binder	Ram extruder	60–95

Derived from Tumuluru et al. (2011)

2.1.3.11 Densification Process Variables

Densification system variables play a significant role in achieving the desired densified biomass properties, such as density and durability. Controlling the densification process variables such as temperature, pressure, and retention time; feedstock properties such as moisture content and particle size; and biomass components such as protein, lignin, starch, wax, and binders impact the quality of the densified products that are produced. In general, preheating of the biomass before densification has a great impact on the quality attributes (such as density and durability) of the densified products. Adding heat to the biomass before densification helps to activate biomass components, such as lignin, as well as reduces densification energy, but results in a high-quality product. Tumuluru (2014) found that preheating corn stover to 110 °C improves pellet quality and reduces the energy consumption of the process. Hall and Hall (1968) found that, for a given moisture content value, the pressure addition of heat in the die reduces energy consumption and improves the durability of the densified biomass. Hill and Pulkinen (1988) found that pellet durability of alfalfa pellets increases by 30–35% when the pelleting temperature is increased from 60 to 104 °C. Sokhansanj et al. (2005) and Mani et al. (2003) observed similar results where they found that increasing the preheating temperature reduces the particle resistances in the die, which in turn increases the quality of the pellets. During densification, increasing the preheating temperature reduces the expansion of the product as the material leaves the pellet or briquette die. Studies conducted by Tumuluru et al. (2016) on woody biomass pelleting indicated that the expansion ratio is influenced by preheating temperature and moisture content. Studies

Table 2.4 Comparison of different densification equipment for biofuels and biopower production

	Extruder	Briquette press	Roller press	Pellet mill	Agglomerator
Pelleting moisture range for woody biomass (%, w.b.)	10–40	10–18	10–15	10–39	40–50
Pelleting moisture range for herbaceous biomass (%, w.b.)	10–40	10–18	10–15	10–38	40–50
Particle size	Smaller (4.76–6.35 mm)	Medium to larger (4.76–19.05 mm)	Medium to larger (4.76–12.07 mm)	Medium to larger (4.76–12.07 mm)	Smaller (135–397 μm)
Wear of contact parts	High	Low	High	High	Low
Output from machine	Continuous	In strokes	Continuous	Continuous	Continuous
Specific energy consumption (kWh/t)	36.8–150	37.4–77	29.91–83.1	16.4–74.5	No information
Throughputs (ton/hr.)	5	2.5	5–10	5	5
Density of briquette	1–1.4 g/cm^3	1–1.2 g/cm^3	0.6–0.7 g/cm^3	0.7–0.8 g/ cm^3	0.4–0.5 g/ cm^3
Maintenance	Low	High	Low	Low	Low
Combustion performance of briquettes	Very good	Moderate	Moderate	Very good	No information
Carbonization of the feedstock during densification	Makes good charcoal	Not possible	Not possible	Not possible	Not possible
Suitability in gasifiers	Suitable	Suitable	Suitable	Suitable	Suitable
Suitability for cofiring	Suitable	Suitable	Suitable	Suitable	Suitable
Suitability for biochemical conversion	Not suitable	Suitable	Suitable	Suitable	No information
Homogeneity of densified biomass	Homogenous	Not homogenous	Not homogeneous	Homogeneous	Homogenous

Derived from Tumuluru et al. (2011), Tumuluru (2014, 2016, 2018b, 2019a), Karunanithy and Muthukumarappan (2010), and Mani (2014)

conducted by Kaliyan and Morey (2006) on understanding Tg temperature found that it is inversely related to moisture content. These authors conducted densification studies using three different preheating temperatures—two that were within the Tg temperature and one that was outside. Results indicated that durability was higher for densified products produced within the Tg temperature as compared to the one completed outside. Similar results were observed by Tumuluru et al. (2017), where it was determined that corn stover moisture content influenced the Tg temperature of biomass components. Studies conducted by Tumuluru (2014, 2016) on the pelleting of herbaceous and woody biomass indicated that higher preheating temperatures of 110 °C produce pellets with higher durability values.

The compressive force applied during pelleting has a great impact on the density and durability of the produced pellets. Butler and McColly (1959) found that the density of alfalfa pellets is directly proportional to the natural logarithm. In general, an increase in applied pressure increases the unit density to the maximum value, yet a further increase does not improve density but may, in fact, have a negative impact on pellet quality. The optimum pressure depends upon die configuration and feedstock type, but going beyond the optimum pressure does not improve density (Ndiema et al. 2002). In general, the application of pressure results in plastic deformation of the biomass particles, thereby influencing the mechanical strength of the densified products. Studies conducted by Yaman et al. (2000) indicated there is an optimum pressure that can influence the plastic deformation of the biomass particles. Feedstock moisture content has a great impact on the pressure necessary to form a densified product. Lower moisture and higher pressure produce a dense pellet, whereas, at high-moisture content, increasing pressure results in moisture loss during extrusion of the pellet out of the die, which reduces pellet density. According to Comoglu (2007) and Denny (2002), the pressure applied in the presence of moisture during densification makes the biomass particles move closer together, which helps to form solid bridges due to chemical modifications. Studies by Demirbas et al. (2004) on the compaction of wastepaper at 7% moisture content showed that an increase in pressure also increased the density. Studies by Li and Liu (2000) suggested that increasing the pressure from 0.24 to 5.0 megapascals per second (MPa/s) increased the density oak sawdust significantly. Holm et al. (2006) suggested that the necessary amount of pelleting pressure depends on the dimensions of the die, such as length and diameter, friction coefficient, and prestressing pressure. These authors also concluded that during pelleting, a back pressure needs to be generated to initiate agglomeration of the particles. The amount of applied pressure has a great impact on the relaxed density of the densified products. Relaxed density also depends on the moisture content of the biomass. Higher moisture content results in higher relaxation of the particles after densification, which can result in the expansion of the product and a reduction in the final density. The final relaxed density of the formed densified product depends on many factors, such as die configuration, compression mode and intensity, and the feedstock properties (Wamukonya and Jenkins 1995). According to Carre et al. (1987) and Miles and Miles (1980), the expansion rate of the densified biomass is higher immediately after pressure removal but will decrease with time until it attains a constant volume.

Shrivastava and Shrivastava (1990) used statistical analysis of rice husks to establish a multiple correlation equation to describe percent volume expansion with respect to die temperature and pressure, as observed in Eq. (2.1).

$$Y = \alpha_0 + \alpha_1 P + \alpha_2 T \tag{2.1}$$

where:

Y = percent volume expansion; die pressure (P) = (kg/m^2); die temperature = T (°C).
α_0, α_1, and α_2 = constants.

The particle residence time in the die has a great impact on the quality of the densified products that are produced (Tabil and Sokhansanj 1996c). Studies conducted by Li and Liu (2000) suggested that retention time has a more significant effect at a lower pressure as compared to that at higher pressures. A 10-second increase in retention times increased log density by 5%, whereas increasing the residence time to greater than 20 s diminished the effect significantly. These results indicated an increase in density up to 40 s of residence time, but had a negligible effect after 40 s. Tumuluru et al. (2020) found that increasing the residence time from 45 to 150 s during pelleting increased the unit density of the corn stover pellets for the two hammermill grind sizes (e.g., 6.35 and 11.11 mm) tested. Studies conducted by Al-Widyan et al. (2002) on olive cake briquetting found that increasing particle residence times from 5 to 20 s did not result in a significant increase in the olive cake briquettes.

The compression ratio of the pellet die is derived from the ratio of the length of the die to the diameter (L/D) of the die. This is also referred to as die geometry and helps to understand the degree of compression the material undergoes in the pellet die. This is an important densification process variable that has a great impact on the energy consumption of the overall process, as well as the energy consumption of the pelleting process. Studies by Butler and McColly (1959) indicated that a higher L/D ratio increased the density and durability of the pellets. Pelleting studies conducted on alfalfa by Hill and Pulkinen (1988) indicated that pellet durability could be increased by about 30–35% by selecting higher L/D ratios of 8–10 and a higher preheating temperature of 104 °C. In general, smaller die diameters result in higher durability values. Heffiner and Pfost (1973) indicated that the quality of the produced pellets using the smallest die had the highest durability values. Studies by Tumuluru et al. (2010b) on pelleting distillers dried grains corroborated with the Heffiner and Pfost (1973) studies where pellets produced using 7.2 mm had lower durability values as compared to pellets produced using 6.4 mm.

2.1.4 Feedstock Properties

Feedstock properties (such as moisture content, particle size, particle shape, and distribution) impact the quality of the produced pellets. Moisture in the biomass in the presence of temperature and pressure activates starch gelatinization, protein denaturation, and fiber solubilization. During compression and extrusion, the starch undergoes gelatinization and protein undergoes denaturation, which supports the binding of the particles (Tumuluru 2018a). The moisture in the biomass up to a certain level acts as a binder by increasing the particle bonding via van der Waal's forces. Studies on the pelleting of spruce wood sawdust at 7–15% (w.b.) resulted in high-strength and dense pellets (Demirbas et al. 2004). Most of the biomasses need about 10% (w.b.) moisture content to make a good quality pellet in terms of density and durability. In the case of alfalfa, the protoplasm acted as a binder at a higher moisture content value of about 19% (w.b.) (Tabil and Sokhansanj 1996c). Biomasses rich in starch can be pelleted at a higher moisture content, as starch in the presence of moisture gelatinizes and helps in binding the particles. On the other hand, biomasses higher in lignin can be pelleted at lower moisture content values. Studies conducted on understanding the compaction characteristics of food powders indicated that increasing moisture content values decreased deformation stresses. In the case of wet powders, they were not deformable as compared to dry powders because they needed more time to reach the preset load (Ollette et al. 1993; Heckel 1961; Peleg and Moreyra 1979). Tumuluru (2014, 2016, 2018a) indicated that good-quality herbaceous and woody biomass pellets could be made in a flat die pellet mill at moisture content values in the range of 28–39% (w.b.).

The other feedstock property that impacts the quality of densified products is grind size. In general, smaller size particle materials results in higher density pellets due to their high-contact surface areas. MacBain (1966) and Payne (1978) indicated that medium to smaller particle size materials is desirable due to their higher surface area and moisture content values, as well as better gelatinization of starch molecules. They also indicated that a certain amount of fine, medium, and coarse grinds is necessary to make high-quality pellets. The particle size should be optimized in terms of production capacity. Going for smaller particle sizes result in high-preprocessing costs, whereas going for bigger particle sizes result in lower quality pellets. Particle size distribution should be focused on reducing preprocessing costs, simultaneously maximizing quality. In their studies on alfalfa pellets, Hill and Pulkinen (1988) indicated that increasing the screen size from 2.8 to 6.4 mm resulted in a reduction in durability by 15%. These same authors concluded that the grind size of the material should be optimized in terms of quality and production costs. Recent studies conducted by Tumuluru et al. (2017) on pelleting corn stover using both a 6.35 and 11.05 mm hammermill grind at 20% (w.b.) moisture content value have indicated that there is about a 50 kg/m^3 of reduction in bulk density, but the durability values of the pellets made with both grind sizes are >97.5%.

2.1.5 Biomass Composition

Biomass composition contributes significantly to the quality of the densified materials. During the densification process, variables like temperature, pressure, and die geometry in the presence of moisture impact the chemical compositions of the biomass. The major reactions that will occur in the biomass are protein-starch reactions, starch gelatinization, protein denaturation, and lignin mobilization. Tumuluru (2018b) described how the various components of biomass change due to pressure, temperature, and moisture content values.

According to Collado and Corke (2003), starch undergoes gelatinization, pasting, and retrogradation reactions in the presence of pressure and temperature. Gelatinization of the starch plays a major role in particle binding during densification. Protein, when heated during densification, undergoes denaturation, which leads to the formation of new bonds and structures with the other components in the biomass (Thomas et al. 1998; Nyanzi and Maga 1992). In his studies on the pelleting of alfalfa, Tabil (1996) indicated that binding improves when there is enough protein in the biomass. The fat in the biomass acts as a lubricant during densification, which reduces friction and increases throughput (Thomas et al. 1998). If the biomass has a higher amount of fat, then binders should be added to produce more durable pellets. Hemicellulose, which is a heteropolysaccharide, has many sugars and other glucose molecules. The amorphous nature of hemicellulose is due to branching, which makes hydrolysis easier in the alkali solution. Hemicellulose can cause natural bonding due to adhesive products that are produced during degradation. Cellulose is an abundant source of carbon in biomass. Zandersons et al. (2004) concluded that the binding in wood-based products mainly depends on converting cellulose to an amorphous state. Hon (1989) indicated that cellulose is a semi-crystalline structure that is bonded by hydrogen, which makes cellulose difficult to dissolve in solvents. Hon (1989) also concluded that cellulose alone is not suitable as an adhesive.

Lignin molecules in plants provide structural support by acting as a glue for the cellulose fibers. Lignin in biomass plays a major role in binding the biomass particles during densification. Above the Tg temperature of >140 °C, lignin exhibits thermosetting properties and acts as an intrinsic resin that helps to produce pellets. Many studies have indicated that moisture in the biomass influences the Tg temperature of lignin. Studies conducted by Lehtikangas (1999) indicated that the lignin molecule softens at about 8–15%, which helps to improve the quality of the pellets. An external binder in very small amounts (e.g., <3% by weight) is added to improve the binding characteristics of lignocellulosic biomass, which have less natural binding components and produce a more durable densified product. Binders also help to reduce wear in production equipment. In pellet production, the addition of external binders is allowed, but they should be clearly specified on the final product. The addition of certain binders results in a corresponding increase in the sulfur content of densified biomass.

2.1.5.1 Densified Biomass Physical Properties

Physical properties (such as density, durability, size, shape, and moisture content values, chemical composition, and energy content) has a major impact on feeding, handling, storage, transportation, and biofuels production. The quality of the densified biomass is dependent on the feedstock type and composition, as well as the densification process variables. It is so important to have densified products to meet quality attributes and standards to retain their quality during storage, transport feeding, and handling.

The final moisture of the densified products, such as the pellets, the briquettes, or the cubes, is generally <10% (w.b.) to make them aerobically stable. Also, lower moisture content values help to reduce losses due to microbial degradation, which results in dry matter losses. In order to maintain lower moisture content values after densification, it is recommended to dry the biomass feedstock to about 10–12% (w.b) before densification. However, the major challenge in doing this is the cost. The final moisture content of the densified product depends on the initial moisture content, temperature, and pressure that are applied. Higher moisture content in the densified products is observed when the initial moisture content of the feedstock is higher. Figures 2.26, 2.27, and 2.28 show how process variables impact the final moisture content of woody and herbaceous biomass pellets when pelleted at high-moisture content values, respectively. Studies conducted by Tumuluru (2014, 2016) on pelleting corn stover and lodgepole pine at higher moisture content values of >28% (w.b) found that the pellet moisture content value was about 15–20% (w.b.), which was dependent on their initial moisture content values. This author also reasoned that the high-moisture pellets could be further dried in low-temperature

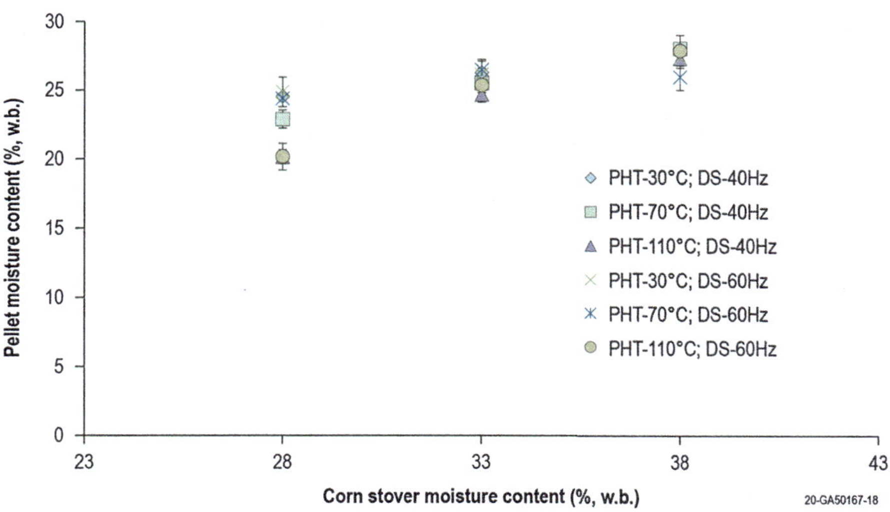

Fig. 2.26 The impact of process conditions on final moisture content values of corn stover pellets made on a flat die pellet mill

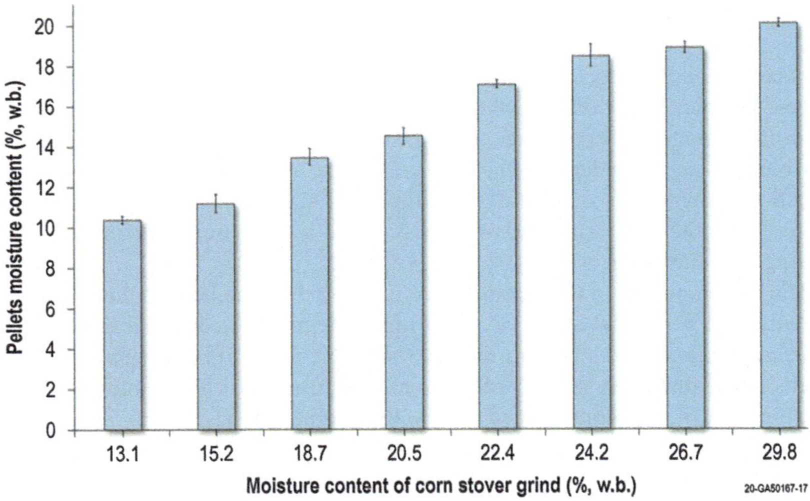

Fig. 2.27 The impact of initial moisture content on the final moisture content of corn stover pellets made in a ring die pellet mill

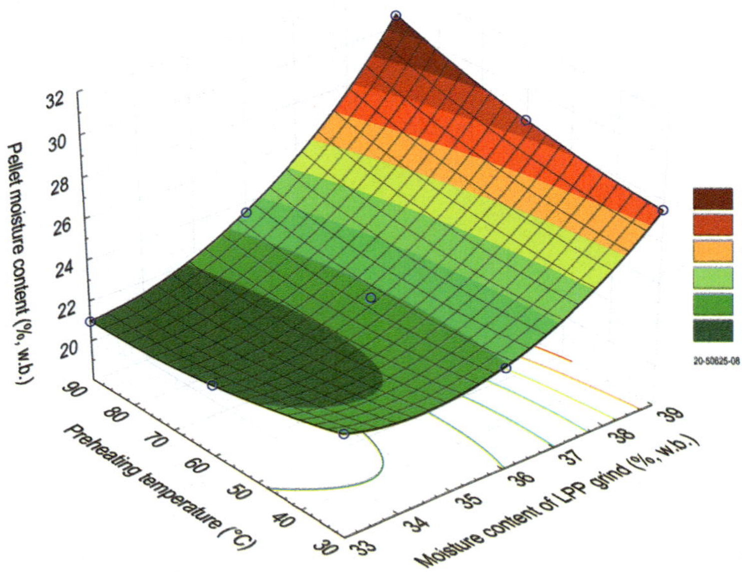

Fig. 2.28 The effect of preheating temperature and moisture content of lodgepole pine (LPP) grind on pellet moisture content

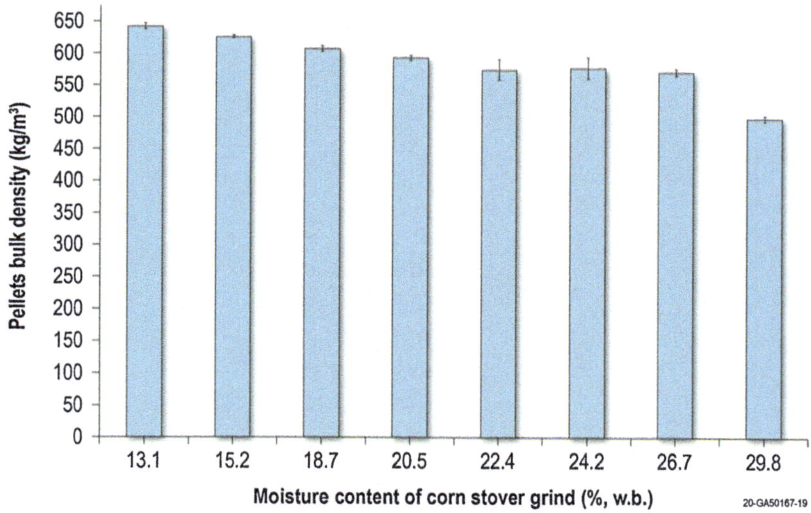

Fig. 2.29 The effect of initial moisture content values on the bulk density of corn stover pellets made in a ring die pellet mill

dryers to <10% (w.b.) in order to keep the pellets aerobically stable. Very low moisture content in the pellets is undesirable as it can result in breakdown issues, fine generation, and the creation of fire issues. In their studies on trying to understand the quality of pellets produced by companies in British Columbia, Canada, Tumuluru et al. (2010b) concluded that moisture content >10% (w.b.) could result in spoilage due to microbial and fungal damage, as well as significant dry matter losses.

The major reason why biomass needs to be densified is that it increases the bulk density of the feedstock by four to six times. The increase in density depends on the type of densification system used. In the case of the pellet mill, the final bulk density that can be reached ranges from 550 to 750 kg/m^3 for woody, herbaceous, and blends of woody and herbaceous biomass pellets (Tumuluru et al. 2010b; Yancey et al. 2013). In the case of briquettes, they reach a maximum density of about 400–500 kg/m^3 (Tumuluru 2019a). An increase in density has a great impact on the storage, handling, and transportation of biomass. According to many researchers, the final density of the biomass feedstock is largely influenced by the initial particle size of the grind, feedstock moisture content, preheating temperature, and applied pressure (Tumuluru 2014, 2016, 2018b; Tumuluru et al. 2016; Mani et al. 2006; Rhen et al. 2005). Figures 2.29, 2.30, and 2.31 show the impact of process conditions on the bulk density of corn stover and woody biomass pellets, respectively. It is clear from these figures that process variables (such as preheating temperature and initial moisture content value) impact bulk density. Studies conducted by Tumuluru et al. (2010b) on the pelleting of distillers dried grains have found that die temperatures and feedstock moisture content value have a significant impact on bulk density. In addition, the densification of biomass not only increases density but also improves

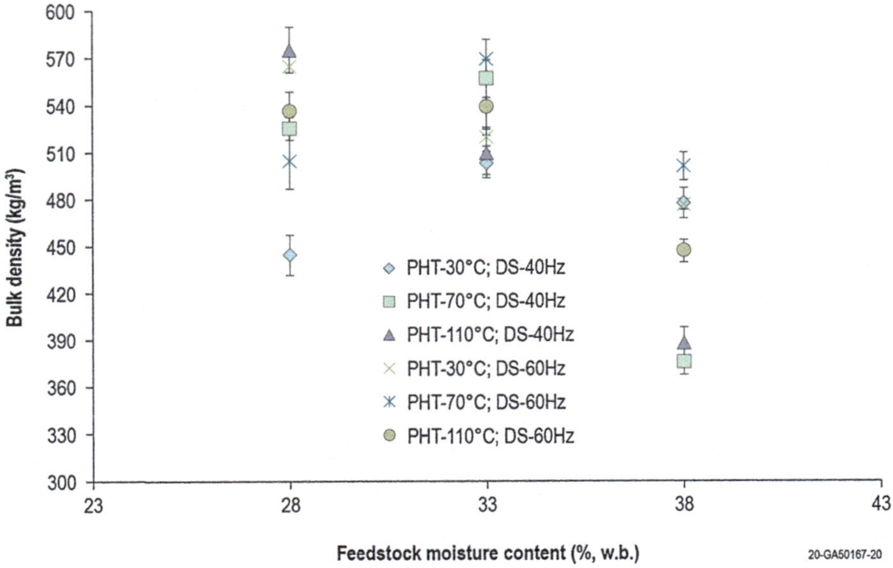

Fig. 2.30 The effect of corn stover moisture content, preheating temperature, and die speed on bulk density in a flat die pellet mill

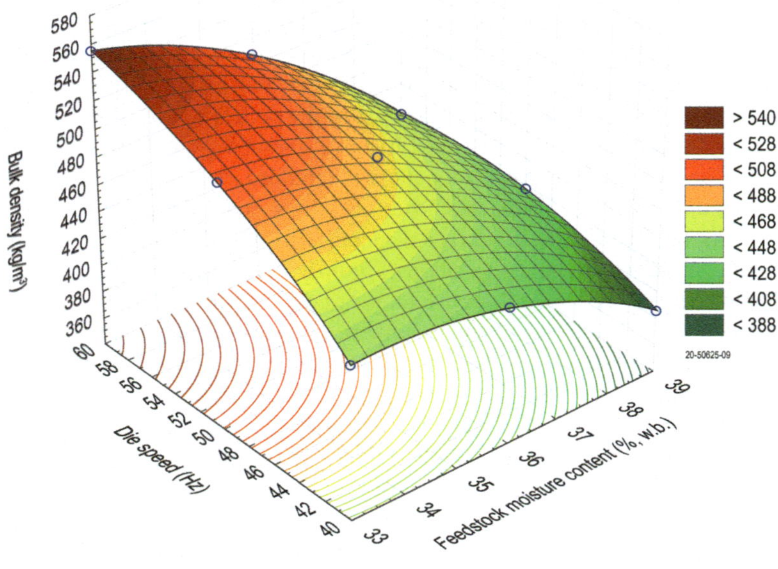

Fig. 2.31 The effect of preheating temperature and moisture content of lodgepole pine (LPP) grind on bulk density

the calorific value (CV). According to Kumar et al. (2009), densified biomass products are preferred over raw material due to its high CV, low-moisture content, and slow-burning properties. It is generally accepted that the grind size of the feedstock has a great impact on density (Mani et al. 2006). Studies conducted by Bergström et al. (2008) indicated there is a marginal impact on density when the particle size distribution of Scots pine sawdust was changed.

The durability of the pellets or other densified products is measured to understand the integrity of pellets during storage, handling, and transportation. Durability or abrasive resistance measurements help to simulate either mechanical or pneumatic handling forces, which help to control pellet quality. In general, moisture in the biomass helps to improve the durability of the densified product. In his studies on pelleting and briquetting, Tumuluru (2019a, b) found that medium- to higher-moisture content values >15% is desirable for improving the durability of the pellets and the briquettes. In addition, when the feedstock is rich in water-soluble starch, the high-moisture content values are desirable because they can act as a binder. Protein in the biomass also helps to improve the durability in the presence of temperature and moisture (Winowiski 1988; Briggs et al. 1999). In most lignocellulosic biomass, lignin plays a major role in the binding of the particles. When pelleting temperatures reach the Tg temperature of lignin, this will help to bind the particles. Typically, at a low-moisture content value of 10%, the Tg temperature of lignin is 140 °C, whereas, at a higher moisture content value, the Tg temperature comes down. In addition, studies conducted by Bradfield and Levi (1984) suggest that higher amounts of lignin and other extractives >36% reduces the durability of the densified products. They reasoned that the auto-adhesive nature of lignin and other extractives decreases due to its excessive gum nature. High amounts of fats in the feedstocks also reduce the durability as it acts as a lubricant rather than a binder. The particle size of the grind has a significant impact on durability. It is generally recommended that medium to bigger particle sizes (e.g., 1/2–1/4-inch) are desirable for briquetting, whereas pelleting particles with a <1/4-inch grind are more desirable. But recent studies on pelleting corn stover at 20% (w.b.) using either a 6.35 or 11.11 mm screen size grind indicated that the durability of the pellets for both grind sizes is >97.5% (Tumuluru et al. 2017), as shown in Fig. 2.32 and Table 2.5. These studies indicate that grind size has more of an impact on bulk density, but not on durability. It is clear from Table 2.5 that increasing the screen size of the stage-2 grinder has impacted bulk density and energy consumption, but not durability. Figure 2.33 indicates how preheating temperature and feedstock moisture content values impact the durability of the wood pellets. It is very clear from Fig. 2.33 that higher preheating temperatures and lower moisture content values improved durability values.

Pellets produced under sub-optimal process conditions (i.e., lower moisture, lower temperature, smaller particle sizes, less compressive force, lower residence times, and less binder) result in less durable pellets. When stored and transported, these pellets result in the generation of fines, which are obviously undesirable. The production of fines results in lost revenue for the pellet producer. The lesser the number of fines, the higher the revenue becomes for the pellet producer. Another major disadvantage of having more fines in pellets is the fire hazards they create. The

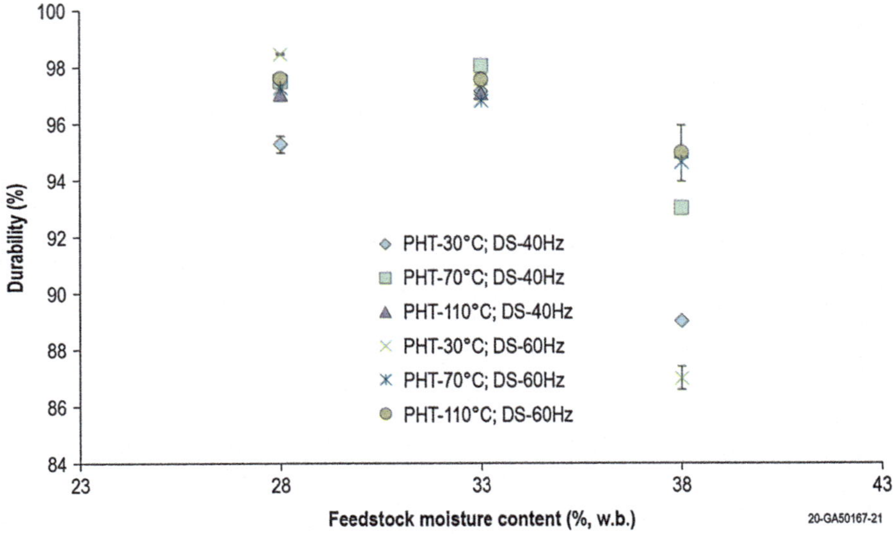

Fig. 2.32 Effect of corn stover moisture content, preheating temperature, and die speed on pellet durability in a flat die pellet mill

Table 2.5 Pellet quality in terms of density and durability for high-moisture pellets made using different hammermill grind sizes

Stage-2 grinder screen size (in.)	Pellet quality				
	Pellet moisture content (%, w.b.)	Bulk density (kg/m^3)	Green durability (%)	Cured durability (%)	Pelleting (kWh/ton)
1/4	13.5	608.4	97.6	97.9	73.78
5/16	15.3	585.5	98.3	98.8	74.97
3/8	15.8	573.6	98.4	98.6	75.73
7/16	15.5	545.5	98.4	98.6	84.34

percentage of fines once they cross threshold levels in terms of particle size and quantity in the storage silos can lead to a dust explosion.

Calorific value (CV) is another important property of biomass feedstock for power generation. Generally, pellets with a higher density have a higher CV as compared to raw biomass. CV is dependent on pelleting process conditions, feedstock properties, and feedstock composition. The typical CVs of pellets are about 18–19 megajoules per kilogram (MJ/kg), according to Tumuluru et al. (2010b). Currently, novel pretreatment technologies, such as torrefaction and hydrothermal carbonization, are being developed with a focus on the improvement of the chemical compositions (e.g., proximate and ultimate properties) and CVs of the biomass.

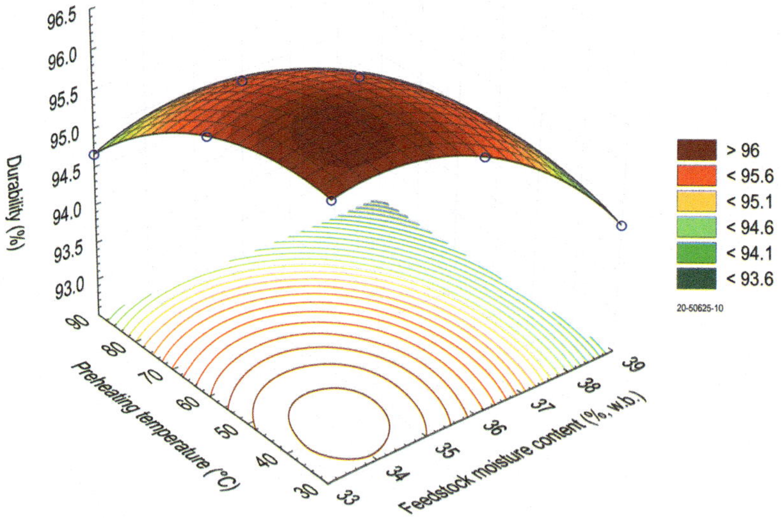

Fig. 2.33 Effect of preheating temperature and moisture content of lodgepole pine (LPP) grind on durability

2.2 Conclusions

There are different densification systems that can be used for producing densified biomass. The most commonly used processes are the pellet mill and the briquette press. During densification, the biomass is subjected to compression pressure, and temperature in the presence of moisture initiates the bonding of biomass particles. Natural binders (such as protein, carbohydrates, and lignin) help the biomass particle to agglomerate. SEM, dynamic analysis, and CT scan analysis indicate that the mobilization of biomass binding components (such as lignin) helps the biomass to form a more densified product. The energy consumption of various densification equipment varies for different densification systems. There are various feedstock properties and densification process variables, which have a great impact on the quality of densified products and energy consumption of the process. Higher moisture reduces density and durability, whereas higher pressures result in higher bulk density products. The major physical properties that have a great impact on feed handling, storage, and transportation are moisture content, bulk density, and durability. The other physical properties (such as fines and CV) impact biofuels conversion processes.

References

Adapa P, Bucko J, Tabil Jr. L, Schoenau G, Sokhansanj S (2002) Pelleting characteristics of fractionated sun-cured and dehydrated alfalfa grinds. ASAE/CSAE North-Central Intersectional Meeting, Saskatoon, Saskatchewan, Canada, 27–28 September 2002

Adapa P, Tabil Jr. L, Schoenau G (2009) Compression characteristics of selected ground agricultural biomass. Ag Eng Int: CIGR Ejournal, Manuscript 1347, 11, June 2009

Al-Widyan MI, Al-Jalil HF, Abu-Zreig MM, Abu-Handeh NH (2002) Physical durability and stability of olive cake briquettes. Can Biosyst Eng 44:341–345

Aqa S, Bhattacharya SC (1992) Densification of preheated sawdust for energy conservation. Energy 17(6):575–578

Bergström D, Israelsson S, Öhman M, Dahlqvist S-A, Gref R, Boman C, Wästerlund I (2008) Effects of raw material particle size distribution on the characteristics of Scots pine sawdust fuel pellets. Fuel Process Technol 89(12):1324–1329

Bilanski WK, Graham VA (1984) A viscoelastic model for forage wafering. T Can Soc Mech Eng 8 (2):70–76

Bradfield J, Levi MP (1984) Effect of species and wood to bark ratio on pelleting of southern woods. For Prod J 34:61–63

Briggs JL, Maier DE, Watkins BA, Behnke KC (1999) Effects of ingredients and processing parameters on pellet quality. Poult Sci 78:1464–1471

Butler JL, McColly HF (1959) Factors affecting the pelleting of hay. Agric Eng 40:442–446

Carre J, Hebert L, Lacrosse L, Schenkel Y (1987) Briquetting agricultural and wood residues: Experience gained with a heated die cylindrical screw press, First FAO/CNRE Workshop on Handling and Processing of Biomass for Energy, 14–15 September 1987, Hamburg, Germany

Collado LS, Corke H (2003) Starch properties and functionalities. In: Kaletunç G, Breslauer KJ (eds) Characterization of Cereals and Flours. Marcel Dekker, Inc., New York, NY, USA, pp 473–506

Comoglu T (2007) An overview of compaction equations. J Fac Pharm Ankara 36(2):123–133

Crowley MM, Zhang F, Repka MA, Thumma S, Upadhye SB, Kumar S, McGinity JW, Martin C (2007) Pharmaceutical application of hot-melt extrusion: Part-1. Drug Dev Ind Pharm 33:909–926

Demirbas A, Şahin-Demirbaş A, Hilal Demirbaş A (2004) Briquetting properties of biomass waste materials. Energ Source 26:83–91

Denny PJ (2002) Compaction equations: a comparison of the Heckel and Kawakita equations. Powder Technol 127:162–172

Gemco Energy (2019) Anyang GEMCO Energy Machinery Co., Ltd. website ©2020. http://www.biofuelmachines.com/. Accessed 8 Apr 2020

Ghebre-Sellassie I (ed) (1989) Mechanism of pellet formation and growth: pharmaceutical pelletization technology. Marcel Dekker, Inc., New York, pp 123–143

Gibson L (2010) Energy tablets, 26 October 2010. Biomass Mag. http://biomassmagazine.com/articles/5086/energy-tablets. Accessed 13 May 2020

Hall GE, Hall CW (1968) Heated-die wafer formation of alfalfa and bermuda grass. Trans ASAE 11:578–581

Harper JM (1979) Food extrusion. CRC Crit Rev Food Sci Nutr 11:155–215

Harper JM (1981) Extrusion of foods, Vol. I and II. CRC Press, Boca Raton, pp 203–212

Heckel RW (1961) An analysis of powder compaction phenomena. T Metall Soc AIME 221:1001–1008

Heffiner LE, Pfost HB (1973) Gelatinization during pelleting. Feedstuffs 45(23):32

Hill B, Pulkinen DA (1988) A study of factors affecting pellet durability and pelleting efficiency in the production of dehydrated alfalfa pellets: A special report. Saskatchewan Dehydrators Association, Tisdale, SK, Canada

Holm JK, Henriksen UB, Hustad JE, Sorensen LH (2006) Toward an understanding of controlling parameters in softwood and hardwood pellets production. Energy Fuel 20:2686–2694

Hon DNS (1989) Cellulosic adhesives. In: Hemmingway RW, Conner AH (eds) Adhesives from Renewable Resources. American Chemical Society, Washington, DC, pp 289–304

Kaliyan N, Morey R (2006) Densification characteristics of corn stover and switchgrass, ASABE Annual International Meeting, 9–12 July 2006, Portland, OR, USA. ASABE Paper No. 066174, ASAE, St. Joseph, MI, USA

Karunanithy C, Muthukumarappan K (2010) Influence of extruder temperature and screw speed on pretreatment of corn stover while varying enzymes and their ratios. Appl Biochem Biotech 162:264–279

Kumar R, Chandrashekar N, Pandey KK (2009) Fuel properties and combustion characteristics of *Lantana Camara* and *Eupatorium* spp. Curr Sci India 97(6):930–935

Kumar RA (2011) Single punch tablet press: current pharmaceutical aspects of industry. Pharmastuff.blogspot.com (https://pharmastuff.blogspot.com/2011/11/single-punch-tablet-press-ppt.html)

Larsson SH, Thyrel M, Geladi P, Lestander TA (2008) High quality biofuels pellet production from precompacted low-density raw materials. Bioresour Technol 99(15):7176–7182

Lehtikangas P (1999) Quality properties of fuel pellets from forest biomass, Ph. D dissertation. The Swedish University of Agricultural Sciences, Uppsala, Sweden

Li Y, Liu H (2000) High pressure densification of wood residues to form an upgraded fuel. Biomass Bioenergy 19:177–186

MacBain R (1966) Pelleting animal feed. Regional Feed School. American Feed Manufacturers Association, Arlington

Mani S (2014) Low cost granulation of lignocellulosic biomass. Final report submitted to the SunGrant Initiative. https://ag.tennessee.edu/sungrant/Pages/SGDOTCompGrants2009-Mani.aspx. Accessed 8 Apr 2020

Mani S, Tabil LG Jr, Sokhansanj S (2003) An overview of compaction of biomass grinds. Powder Handl Process 15(3):160–168

Mani S, Tabil LG Jr, Sokhansanj S (2004) Evaluation of compaction equations applied to four biomass species. Can Biosyst Eng 46:3.55–3.61

Mani S, Tabil LG Jr, Sokhansanj S (2006) Specific energy requirement for compacting corn stover. Bioresour Technol 97:1420–1426

Mewes E (1959) Berechung der druckverteilung an stroh-und heupressen (Calculation of the pressure distribution in straw and hay balers). Landtechnische Forschung 9(6):160–170

Miles TR, Miles TR Jr (1980) Densification systems for agricultural residues. In: Jones JL, Radding SB (eds) Thermal Conversion of Solid Wastes and Biomass. American Chemical Society, Washington, DC, pp 179–191

Ndiema CKW, Manga PN, Ruttoh CR (2002) Influence of die pressure on relaxation characteristics of briquetted biomass. Energ Convers Manage 43:2157–2161

Nyanzi FA, Maga JA (1992) Effect of processing temperature on detergent-solubilized protein in extrusion-cooked cornstarch/soy protein subunit blends. J Agric Food Chem 40:131–133

O'Dogherty MJ, Wheeler JA (1984) Compression of straw to high densities in closed cylindrical dies. J Agric Eng Res 29(1):61–72

Ollett AL, Kirby AR, Parker R, Smith AC (1993) A comparative study of the effects of water content on the compaction behavior of some food materials. Powder Technol 75:59–65

Payne JD (1978) Improving quality of pellet feeds. Milling Feed Fert 161:34–41

Peleg M, Moreyra R (1979) Effect of moisture on the stress relaxation pattern of compacted powders. Powder Technol 23:277–279

Pietsch WB (1984) Size enlargement methods and equipments—Part 2: Agglomerate bonding and strength. In: Fayed ME, Otten L (eds) Handbook of Powder Science and Technology. Van Nostrand Reinhold Co., New York, NY, USA, pp 231–252

Rhen C, Gref R, Sjöström M, Wästerlund I (2005) Effects of raw material moisture content, densification pressure, and temperature on some properties of Norway spruce pellets. Fuel Process Technol 87(1):11–16

Rumpf H (1962) The strength of granules and agglomerates. In: Knepper WA (ed) Agglomeration. Interscience Publishers, New York, NY, USA, pp 379–418

Sastry KVS, Fuerstenau DW (1973) Mechanisms of agglomerate growth in green pelletization. Powder Technol 7:97–105

Schineberger GL (1971) Understanding adhesives. Hitchcock Publishing Co., Wheaton, pp 3.6–3.7

Sherrington PJ, Oliver R (1981) Fundamentals of particle size enlargement by granulation. In: Sherrington PJ, Oliver R (eds) Granulation. Heyden and Sons Ltd., London, England, pp 7–54

Shrivastava M, Shrivastava P (1990) Briquetting of rice husk under hot compression. In: Salokhe VM, Ilangantileke SG (eds) Proceedings of the International Agricultural Engineering Conference and Exhibition, Bangkok, Thailand, 3–6 December 1990, pp 663–672

Sokhansanj S, Mani S, Bi X, Zaini P, Tabil Jr. LG (2005) Binderless pelletization of biomass, ASAE Annual International Meeting, 17–20 July 2005, Tampa, FL, USA. ASAE paper no. 056061. ASAE, St. Joseph, MI, USA

Tabil Jr. LG (1996) Binding and pelleting characteristics of alfalfa, Unpublished Ph.D. Thesis, Department of Agricultural and Bioresource Engineering, University of Saskatchewan, Saskatoon, Saskatchewan, Canada

Tabil LG Jr, Sokhansanj S (1996a) Compression and compaction behavior of alfalfa grinds: Part 1: Compression behavior. Powder Handl Process 8(1):17–23

Tabil LG Jr, Sokhansanj S (1996b) Compression and compaction behavior of alfalfa grinds: Part 2: Compaction behavior. Powder Handl Process 8(2):117–122

Tabil LG Jr, Sokhansanj S (1996c) Process conditions affecting the physical quality of alfalfa pellets. Appl Eng Agric 12(3):345–350

Thomas M, van Vliet T, van der Poel AFB (1998) Physical quality of pelleted animal feed 3: Contribution of feedstuff components. Anim Feed Sci Tech 70:59–78

Tumuluru JS (2014) Effect of process variables on the density and durability of the pellets made from high moisture corn stover. Biosyst Eng 119:44–57

Tumuluru JS (2015) High moisture corn stover pelleting in a flat die pellet mill fitted with a 6 mm die: Physical properties and specific energy consumption. Energy Sci Eng 3(4):327–341

Tumuluru JS (2016) Specific energy consumption and quality of wood pellets produced using high moisture lodgepole pine grind in a flat die pellet mill. Chem Eng Res Des 110:82–97

Tumuluru JS (2018a) Why biomass preprocessing and pretreatments? In: Tumuluru JS (ed) Biomass Preprocessing and Pretreatments for Production of Biofuels. CRC Press, Boca Raton, FL, USA, pp 1–14

Tumuluru JS (2018b) Effect of pellet die diameter on density and durability of pellets made from high moisture woody and herbaceous biomass. Carbon Resour Convers 1(1):44–54

Tumuluru JS (2019a) Effect of moisture content and hammer mill screen size on the briquetting characteristics of woody and herbaceous biomass. KONA Powder Part J 36:241–251

Tumuluru JS (2019b) Pelleting of pine and switchgrass blends: effect of process variables and blend ratio on the pellet quality and energy consumption. Energies 12(7):1198

Tumuluru JS, Yancey N (2018) Conventional and advanced mechanical preprocessing methods for biomass: performance quality attributes and cost analysis. In: Tumuluru JS (ed) Biomass Preprocessing and Pretreatments for Production of Biofuels. CRC Press, Boca Raton, FL, USA, pp 15–49

Tumuluru JS, Wright CT, Kenney KL, Hess JR (2010a) A review on biomass densification technologies for energy applications. Idaho National Laboratory Technical Report INL/EXT-10-18420, Idaho Falls, Idaho, USA. http://www.inl.gov/bioenergy. Accessed 8 Apr 2020

Tumuluru JS, Tabil Jr. LG, Opoku A, Mosqueda MR, Fadeyi O (2010b) Effect of process variables on the quality characteristics of pelleted wheat distiller's dried grains with soluble. Biosyst Eng 105(4):466–475

Tumuluru JS, Wright CT, Hess JR, Kenney KL (2011) A review of biomass densification systems to develop uniform feedstock commodities for bioenergy application. Biofuels Bioprod Biorefin 5(6):683–707

Tumuluru JS, Conner CC, Hoover AN (2016) Method to produce durable pellets at lower energy consumption using high moisture corn stover and a corn starch binder in a flat die pellet mill. J Vis Exp 112:e54092

Tumuluru JS, Yancey N, McCulloch R, Fox C, Conner CC, Hartley D, Dee M, Plummer M (2017) Biomass engineering: size reduction, drying, and densification of high moisture biomass. Feedstock supply and logistics platform, 2017 project peer review, US Department of Energy (DOE) Bioenergy Technologies Office (BETO), 7 March 2017, Denver, CO, USA. https://www.energy.gov/sites/prod/files/2017/05/f34/fsl_tumuluru_1222.pdf. Accessed 8 Apr 2020

Tumuluru JS, Fillerup E, Kane J, Murray DJ (2019) Biomass engineering: size reduction, drying, and densification of high moisture biomass, feedstock supply and logistics platform, US Department of Energy, Project Peer Review, DOE Bioenergy Technologies Office, 6 March 2019, Denver, CO, USA. https://www.energy.gov/sites/prod/files/2019/04/f61/Size%20Reduction%2C%20Drying%20and%20Densification%20of%20High%20Moisture%20Biomass_NL0026654.pdf. Accessed 8 Apr 2020

Tumuluru JS, Fillerup E, Kane J, Murray DJ (2020) Advanced imaging techniques to understand the impact of process variables on the particle morphology in a corn stover pellet. Chem Eng Res Des 161:130–145

Wamukonya L, Jenkins B (1995) Durability and relaxation of sawdust and wheat-straw briquettes as possible fuels for Kenya. Biomass Bioenergy 8(3):175–179

Winowiski T (1988) Wheat and pellet quality. Feed Manage 39:58–64

Wood JF (1987) The functional properties of feed raw materials and the effect on the production and quality of feed pellets. Anim Feed Sci Tech 18:1–17

Yaman S, Şahan M, Haykiri-açma H, Şeşen K, Küçükbayrak S (2000) Production of fuel briquettes from olive refuse and paper mill waste. Fuel Process Technol 68:23–31

Yancey NA, Tumuluru JS, Wright CT (2013) Drying, grinding, and pelletization studies on raw and formulated biomass feedstocks for bioenergy applications. J Biobaased Mater Bioenergy 7:549–558

Yehia KA (2007) Estimation of roll press design parameters based on the assessment of a particular nip region. Powder Technol 177:148–153

York P, Pilpel N (1972) The effect of temperature on the mechanical properties of some pharmaceutical powders in relation to tableting. J Pharm Pharmacol 24:47P–56P

Zandersons J, Gravitis J, Zhurinsh A, Kokorevics A, Kallavus U, Suzuki CK (2004) Carbon materials obtained from self-binding sugar cane bagasse and deciduous wood residues plastics. Biomass Bioenergy 26:345–360

Chapter 3
Densification Process Models and Optimization

Abstract Compression tests and models are commonly used to understand the densification characteristics of metal, pharmaceuticals, agriculture, and food. Compression models help reveal the behavior of biomass grinds or particles during pelleting and can help to optimize the pressures required to manufacture a quality pellet. Pellets are formed by using either pressure agglomeration or compaction processes in which the particles are bound together with or without binders. The compression models expressing the relationship between pressure and time during the compaction of different raw materials are available in the literature. This chapter will focus on the various compression models and their suitability for different biomass feedstocks. This chapter also focus on how response surface methodology and advanced computational methods can be used for densification process modeling and optimization.

Keywords Compression process models · Process variables · Product properties · Response surface methodology · Optimization · Evolutionary algorithms

3.1 Introduction

The compaction characteristics of ground biomass depend on the mechanical properties of the solid particles in their makeup. The process variables that impact these compression characteristics are moisture, pressure, and residence time of the particle in the pellet die. Fundamental understanding of the compression process is necessary to design the densification system and to produce a densified product with the desired quality attributes such as density. The compaction behavior of the powder varies based on the power type. In general, the quality of the densified product depends on the physical forces necessary to help the particles bind together (Tabil and Sokhansanj 1996). According to Adapa et al. (2002), the three physical forces that impact densified biomass production are thermal, mechanical, and atomic. There are three different stages in compaction. In the first, the particles rearrange themselves to form a closely packed mass. In this stage, the particle retains most of its original material properties, even though the energy is dissipated into the particles

© Springer Nature Switzerland AG 2020
J. S. Tumuluru, *Biomass Densification*,
https://doi.org/10.1007/978-3-030-62888-8_3

due to interparticle and particle-to-wall friction. As the compaction pressure increases in the second stage, the particles undergo elastic and plastic deformation as they are pressed up against each other. According to Rumpf (1962) and Sastry and Fuerstenau (1973), an increase in the interparticle contact area results in bonding due to van der Waal's forces. At this stage, the particles that have not undergone elastic and plastic deformation become brittle and break, which also contributes to mechanical interlocking (Gray 1968). Where the compaction pressures are further increased during the third and final stage, the pores inside the particles are broken, and the density of the compacted biomass approaches the true or solid densities of its component ingredients. If a eutectic mixture is formed at this point, then the heat generated at the contact points can result in local material melting. Once the molten material is cooled, solid bridges are formed. The various components that exist in the biomass are lignin, cellulose, hemicellulose, protein, fiber, and wax. Tumuluru (2018) discussed how these components interact during the application of pressure and temperature in the presence of moisture. Among these components, lignin—which is available in large quantities in biomass—has a glass transition temperature of about 140 °C. When biomass is heated in the die, the lignin softens, melts, and exhibits setting properties (van Dam et al. 2004). Tabil and Sokhansanj (1996) observed a similar compaction mechanism during the pelleting of alfalfa. Studies conducted by Alderborn and Wikberg (1996) on the compaction of pharmaceutical powders observed mechanisms such as particle rearrangement, deformation, densification, fragmentation, and attrition. Additional studies conducted by Johansson et al. (1995) and Johansson and Alderborn (1996) on the microcrystalline behavior during pharmaceutical table formation indicated there are two types of mechanisms—permanent deformation (change in the shape of individual particles) and densification (porosity decrease of the individual compacts). According to Sonnergaard (2001), the compression of the particle occurs in different stages; it is challenging to develop a mono-variate equation for the entire range.

3.1.1 Compression Models

Walker (1923), Heckel (1961), Cooper and Eaton (1962), and Kawakita and Lüdde (1971) proposed several models to understand the compaction behavior of powders. Denny (2002) and Mani et al. (2003) reviewed the application of these models for pharmaceutical and biomass powders. In pharmaceutical compaction studies, the Heckel and Cooper-Eaton models are widely used, whereas, for soft and fluffy material, the Kawakita and Lüdde model is widely used. Studies conducted by Tabil and Sokhansanj (1996) reviewed the application of different models for alfalfa pellet production. They concluded that the Heckel, Kawakita and Lüdde, and Cooper-Eaton models all fit the alfalfa compression data well.

3.1.1.1 Spencer and Heckel Model

Spencer and Heckel (Heckel 1961) developed an equation to relate the density in terms of packing fractions as a function of applied pressure:

$$\ln \frac{1}{1-\rho_f} = m\rho + b \tag{3.1}$$

where:

$$\rho_f = \frac{\rho}{\rho_1 x_1 + \rho_2 x_2}. \tag{3.2}$$

Equations (3.1) and (3.2) are used to describe the compression behavior of the powder materials where the two stages of compression are described by the constants m and b—pre-occupation and particle rearrangement—due to densification.

Shivanand and Sprockel (1992) indicated that constant b is related to the relative density at particle rearrangement (ρ_f) by Eq. (3.3):

$$b = \ln \frac{1}{1-\rho_f}. \tag{3.3}$$

Based on Eq. (3.3), higher ρ_f results in higher volume reduction due to more particle rearrangement, whereas m is the mean yield pressure required to induce elastic deformation. Higher m values higher plastic deformation due to low yield pressure, which can be an indicator that the material is easy to compress.

3.1.1.2 Walker Model

A model was developed by Walker (1923) based on the experimental data on the compressibility of powders and expressed the volume ratio (V_R) as a function of applied pressure (P):

$$V_R = m \ln P + b \tag{3.4}$$

$$V_R = \frac{V}{V_s} \tag{3.5}$$

where:

P = applied pressure per megapascal (MPa)
V_R = volume ratio
V = volume of the compact at pressure P (m^3)
V_s = solid material that is free of the void (m^3)

3.1.1.3 Jones Model

A density- and pressure-based model was developed by Jones (1960) for compacted metal powder:

$$\ln \rho = m \ln P + b \tag{3.6}$$

$$b = \ln\left(\frac{1}{1-\rho_0}\right) \text{ and } \rho f = \frac{\rho}{\rho_1 x_1 + \rho_2 x_2} \tag{3.7}$$

where:

ρf = material relative density after particle rearrangement
ρ_0 = powder mixture relative density (kg/m^3)
ρ_1 and ρ_2 = component of the mixture particle density (kg/m^3)
x_1 and x_2 = component of the mixture mass fraction

Constants b and m are determined based on the intercept and slope of the extrapolated linear region of the plot $\ln\left(\frac{1}{1-\rho f}\right)$ vs. P.

Higher values of ρf will indicate increased volume reduction of the samples due to particle rearrangement.

Constant m is shown as reciprocal to the mean yield pressure required to induce the elastic deformation (York and Pilpel 1972). A higher m value will indicate that yield pressure is low and plastic deformation onsets at relatively low pressures, showing that the material is more compressible.

3.1.1.4 Cooper-Eaton Model

The Cooper-Eaton model (Cooper and Eaton 1962) assumes compression as nearly two independent probabilistic processes: (1) filling of voids of the same size as the particles and (2) filling of voids smaller than the particles:

$$\frac{V_0 - V}{V_0 - V_s} = a_1 e^{-\frac{k_1}{P}} + a_2 e^{-\frac{k_2}{P}} \tag{3.8}$$

where:

V_0 = volume of the compact at zero pressure (m^3), a_1, a_2
k_1, k_2 = Constants for Cooper-Eaton model

Understanding the physical significance of the constants in the equation is the practical limitation of the model (Cooper and Eaton 1962).

According to Comoglu (2007), it is more suitable for single-component systems.

3.1.1.5 Kawakita and Lüdde Model

The Kawakita and Lüdde model (Kawakita and Lüdde 1971) includes the pressure and volume factors:

$$\frac{P}{C} = \frac{1}{ab} + \frac{P}{a} \tag{3.9}$$

$$C = \frac{V_0 - V}{V_0} \tag{3.10}$$

where:

C = degree of volume reduction or engineering strain
a and b = Kawakita-Lüdde model constants related to characteristics of the powder
$\frac{P}{C}$ and P values can be evaluated graphically.

According to Denny (2002) and Kawakita and Lüdde (1971), the equation holds true for soft and fluffy powders. The fluctuations in the measured value of V_0 can result in any deviations from this expression. A study conducted by Mani et al. (2004) indicated that constant a is equal to the initial porosity of the sample, whereas the constant $\frac{1}{b}$ is due to failure stress in the case of piston compression.

3.1.1.6 Sonnergaard Model (Log-Exp-Equation)

A log-exp equation that considers a logarithmic decrease in volume by fragmentation and an exponential decay representing the plastic deformation of powders was developed by Sonnergaard (2001).

$$V = V_1 - w \log P + V_0 \exp\left(-P/P_m\right) \tag{3.11}$$

where:

V_1 = volume at pressure 1 (MPa)
P_m = mean pressure (MPa)
w = model constant

Sonnergaard (2001) suggests that his model provides better regression values than either the Cooper-Eaton (Cooper and Eaton 1962) or the Kawakita and Lüdde (1971) models. In general, this model suits the best for medium pressure applications (approximately 50 MPa).

3.1.1.7 Panelli-Filho Model

Equation 3.12 gives the model developed by Panelli-Filho (2001):

$$\ln \frac{1}{1-\rho_r} = A\sqrt{P} + B \tag{3.12}$$

where:

ρ_r = relative density of the compact
A = particle deformation parameter related to densification
B = powder density parameter at the start of compression

Adapa et al. (2009) studied the compression characteristics of wheat, oat, canola, and barley straws at 10% (w.b.) moisture content. Tables 3.1, 3.2, 3.3, 3.4, and 3.5 indicate the compression models constants for barley, canola, oat, and wheat straw using the Jones, Heckel, Cooper-Eaton, Kawakita-Lüdde, and Panelli-Filho models, respectively. Their results indicated that the Kawakita-Lüdde model provided an excellent fit with the R^2 values of 0.99 for all the selected straws. The models also indicated that the oat and the canola had the highest failure stress and porosity, respectively. The Cooper-Eaton model indicated that the wheat straw samples got

Table 3.1 Jones model constants for barley, canola, oat, and wheat straw

	Constants		
Feedstock	m	b	R^2 value
Barley straw	0.052	6.643	0.42
Canola straw	0.138	6.243	0.71
Oat straw	0.120	6.333	0.62
Wheat straw	0.089	6.418	0.47

Table 3.2 Heckel model constants for barley, canola, oat, and wheat straw

	Constants		
Feedstock	m	b	R^2 value
Barley straw	0.001	0.959	0.31
Canola straw	0.002	0.724	0.70
Oat straw	0.002	0.771	0.55
Wheat straw	0.001	0.740	0.34

Table 3.3 Cooper-Eaton model constants for barley, canola, oat, and wheat straw

	Constants		R^2 value		
Feedstock	a1	a2	k1	k2	R^2 value
Barley straw	0.7025	0.2000	1.3025	1.3024	0.52
Canola straw	1.8141	−0.9117	−1.6542	−6.0377	0.72
Oat straw	0.8958	0.0202	4.5734	−31.7373	0.64
Wheat straw	1.4503	−0.6091	−11.3827	−23.3329	0.64

Table 3.4 Kawakita-Lüdde model constants for barley, canola, oat, and wheat straw

Feedstock	Constants		R^2 value
	a	1/b	
Barley straw	0.738	0.776	0.99
Canola straw	0.749	3.801	0.99
Oat straw	0.751	3.149	0.99
Wheat straw	0.720	1.727	0.99

Table 3.5 Panelli-Filho model constants for barley, canola, oat, and wheat straw

Feedstock	Constants		R^2 value
	A	B	
Barley straw	0.020	0.866	0.36
Canola straw	0.045	0.530	0.74
Oat straw	0.046	0.576	0.57
Wheat straw	0.023	0.634	0.40

easily compacted, but according to the Jones model, the canola and oat straws compressed more easily as compared to the barley and wheat straws.

Mani et al. (2004) studied the compression characteristics of wheat, straw, barley straw, corn stover, and switchgrass. These authors also varied the hammermill screen size and moisture content of each of the feedstocks. The Heckel, Copper-Eaton, and Kawakita-Lüdde models were the ones studied, as shown in Table 3.6. Mani et al. (2004) concluded that based on the Cooper-Eaton model parameters, the dominant compaction mechanism for the biomass tested was the rearrangement of the particles followed by elastic and plastic deformation, but the mechanical interlocking mechanism was found to be negligible. The Kawakita-Lüdde model predicted the highest and lowest yield strength for the corn stover. Fig. 3.1 indicates that the Kawakita-Lüdde model fits best for the agricultural biomass—such as the wheat straw, barley straw, corn stover, and switchgrass—while Fig. 3.2 shows how the density changes with alterations in applied pressure for these feedstocks (Adapa et al. 2009).

3.1.1.8 Holm's Model

The fundamental relationship between the length to diameter (L/D) ratio, extrusion pressure, preheating temperature, and feedstock moisture was used to develop the scale-up methodology. Holm et al. (2006) showed that by balancing the frictional forces on the pellet/die interface with the compression forces acting on the top of the pellet, the relation of pressure to the L/D ratio could be expressed as:

$$P(x) = \frac{P_{No}}{\nu_{LR}}\left(\exp\left(4\mu\nu_{LR}x\right) - 1\right) \tag{3.13}$$

where x is the ratio of the L/D, $P(x)$ is the extrusion pressure, P_{No} is the prestressing pressure required to initiate pelleting, ν_{LR} is Poisson's ratio for radial strain given lateral stress, and μ is the kinetic coefficient of friction between the pellet and the die.

Table 3.6 Heckel, Copper-Eaton, and Kawakita-Lüdde models for wheat straw, barley straw, corn stover, and switchgrass

Model parameters	Hammermill screen size (mm)	Moisture content (%, w.b.) Wheat straw		Barley straw		Corn stover		Switchgrass	
		12	15	12	15	12	15	12	15
Pr (MPa)	3.2	2.0	1.7	0.8	1.0	1.2	0.7	3.3	4.8
	1.6	0.8	1.7	1.3	0.6	0.8	0.5	2.3	4.8
	0.8	0.4	0.6	3.1	1.9	0.9	0.6	5.5	5.1
Pd (MPa)	3.2	1.1	1.0	0.8	1.0	1.2	0.7	3.3	2.0
	1.6	2.0	1.7	1.8	2.0	0.8	0.5	2.0	2.0
	0.8	2.0	0.6	2.0	1.9	0.9	0.6	2.0	2.0
a1 + a2	3.2	1.01	1.0	1.0	0.99	1.01	0.99	1.03	1.03
	1.6	0.9	0.98	0.99	0.97	0.99	0.98	0.99	0.98
	0.8	0.97	0.95	0.99	0.98	0.98	0.97	1.0	0.97
A	3.2	0.91	0.91	0.91	0.90	0.89	0.88	0.91	0.91
	1.6	0.91	0.89	0.91	0.89	0.87	0.86	0.85	0.85
	0.8	0.89	0.86	0.91	0.89	0.87	0.86	0.85	0.82
a*	3.2	0.91	–	0.91	–	0.89	–	0.88	–
	1.6	0.92	–	0.91	–	0.88	–	0.86	–
	0.8	0.91	–	0.91	–	0.89	–	0.85	–
1/b (MPa)	3.2	1.64	1.60	0.71	1.07	1.09	0.63	3.65	3.92
	1.6	1.71	1.15	1.78	1.68	0.59	0.44	2.04	4.03
	0.8	1.29	1.32	3.05	1.70	0.75	0.6	3.97	4.03

*Indicates theoretical initial porosity of biomass grinds

Material properties and the prestressing pressure term are determined experimentally using single pelleting tests conducted with an Instron model 5982 and a 100 kN load cell. Initially, a small amount of material is compressed against a backstop. Material is added and compressed until the desired length—and thus the L/D ratio—is achieved, at which point the backstop is removed, and the maximum force required to initiate motion is recorded.

Changing the L/D ratio affects extrusion pressure and, in turn, pellet density. Stelte et al. (2012) considered pellet properties and scale-up based on changing the L/D ratio up to five. These authors proved that a gap exists between their experimental data and an industrial scale pellet press, as shown in Fig. 3.3. In this work, we have developed experimental data for a wide L/D ratio, ranging from two up to nine.

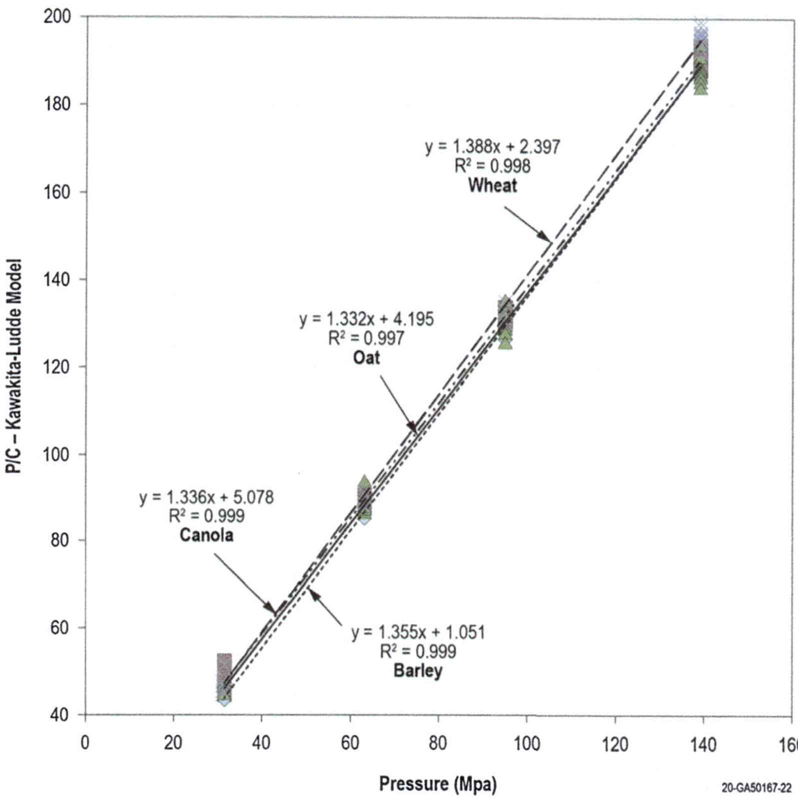

Fig. 3.1 The Kawakita-Lüdde model matched up with the experimental data obtained from the densification of selected agricultural biomass (derived from Adapa et al. 2009)

3.1.2 Application of the Compression Models for Process Scale-Up

3.1.2.1 Scale-Up Methodology

This work combines Holm's extrusion pressure model with the Kawakita-Lüdde density model under the assertion by Stelte et al. (2012) that for steady-state operation, the extrusion pressure must be equal to the compaction pressure in the die channel. The coupling of these two models is illustrated in Fig. 3.4, where step 1 defines the L/D ratio, step 2 determines the extrusion pressure for the L/D ratio, and step 3 determines the density at the compaction pressure from step 2 (Tumuluru et al. 2017).

An experimental plan was developed to quantify the impact of different preheating temperatures, moisture content values, and L/D ratios (see Table 3.7). For the Holm model, the independent variables were preheating temperature,

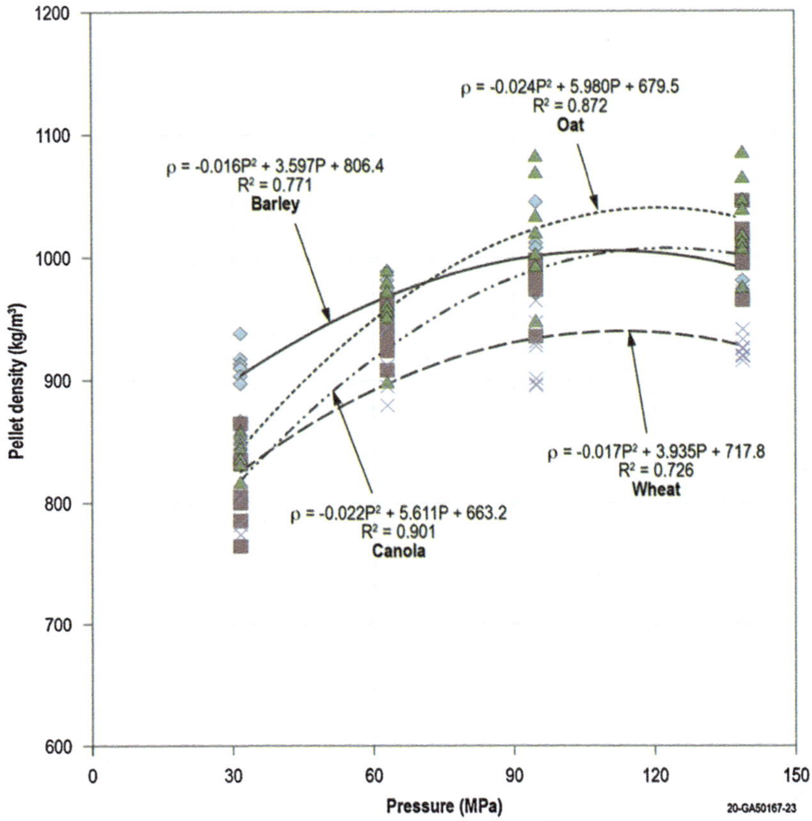

Fig. 3.2 The density of the pellets and their empirical equations for four agricultural straw samples at four pressure levels

moisture content values, and L/D ratio, while the dependent variable was extrusion pressure. For the Kawakita-Lüdde model, the independent variables were preheating temperature, moisture content values, and compaction pressure, while the dependent variable was unit density. Fig. 3.5 indicates the single pellet press used in the present study (Tumuluru et al. 2017).

Typical model constants for the Holm and Kawakita-Lüdde models for corn stover at 110 °C and moisture contents between 20 and 30% (w.b.) are given in Tables 3.8 and 3.9, respectively. The coefficient of determination values indicates that the models have adequately described the experimental data. Similar model constants were obtained for other preheating temperatures and feedstock moisture contents based on Table 3.7 (Tumuluru et al. 2017).

Figure 3.6 shows the pellets produced for a typical Holm model experiment for corn stover at 110 °C preheating temperature for various L/D ratios and moisture content values between 20 and 30% (w.b.), for a constant compaction pressure of

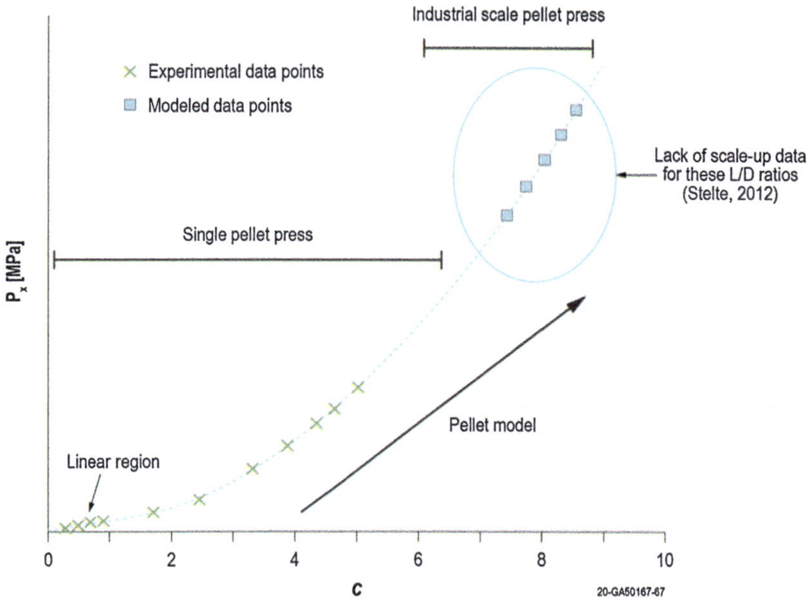

Fig. 3.3 Qualitative description showing how Holm's model can be used for scale-up (derived from Stelte et al. 2012)

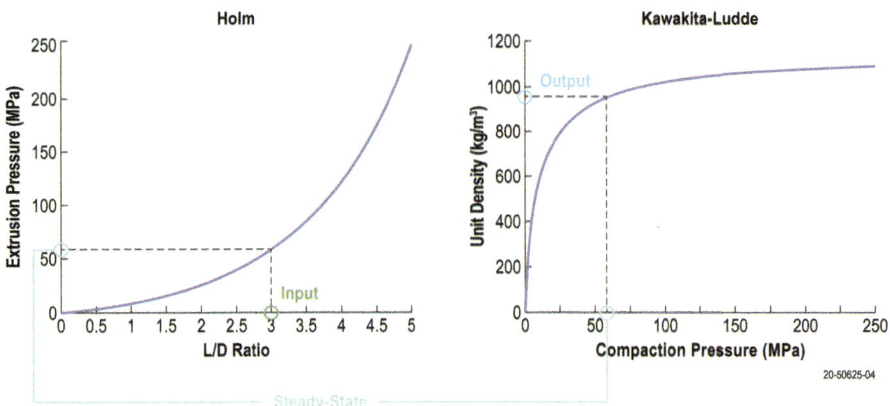

Fig. 3.4 Output and coupling between the Holm and Kawakita-Lüdde models

190 MPa. Figure 3.7 shows typical pellets produced for the Kawakita-Lüdde experiment for corn stover at 110 °C preheating temperature at different feedstock moisture content values. A family of curves was constructed for different combinations of moisture content values, preheating temperatures, and L/D ratios based on Table 3.9. A typical family of curves for 110 °C is shown in Fig. 3.8 for corn stover.

Table 3.7 Experimental plans for both the Holm and Kawakita-Lüdde models

Model	Independent variable	Range	Dependent variable
Holm	Preheating temperature	70–110 °C	Extrusion pressure (MPa)
	Moisture content	20–30% w.b.	
	L/D ratio	1–8	
Kawakita-Lüdde	Preheating temperature	70–110 °C	Unit density (kg/m^3)
	Moisture content	20–30% w.b.	
	Compaction pressure	30–170 MPa	

Derived from Tumuluru et al. (2017)

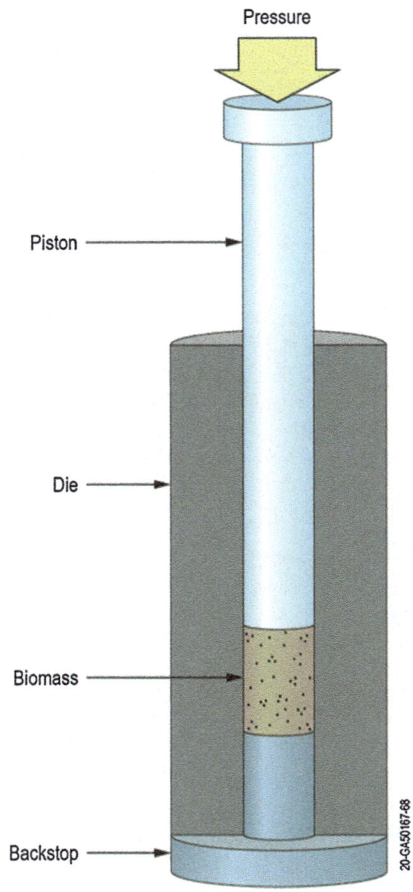

Fig. 3.5 Single pellet press used for generating data using Holm and Kawakita-Lüdde model

These curves help to predict the L/D ratios and extrusion pressures necessary to produce a pellet with a certain unit density at a given feedstock moisture content value. The single pelleting studies indicate that the die dimensions (e.g., L/D ratio) govern the pressure—and in turn, the density—of the produced pellet. The Holm and

Table 3.8 Empirical constants for Holm's model using corn stover

Temperature (°C)	Moisture content (% w.b.)	P_{No}	ν_{LR}	μ	R^2
110	30	0.676	0.167	0.730	0.99
	25	0.669	0.161	0.660	0.96
	20	0.281	0.203	0.715	0.99

Table 3.9 Empirical constants for Kawakita-Lüdde's model using corn stover

Temperature (°C)	Moisture content (% w.b.)	a	b	R^2
110	30	0.905	1.183	0.85
	25	0.913	1.277	0.99
	20	0.892	0.855	0.97

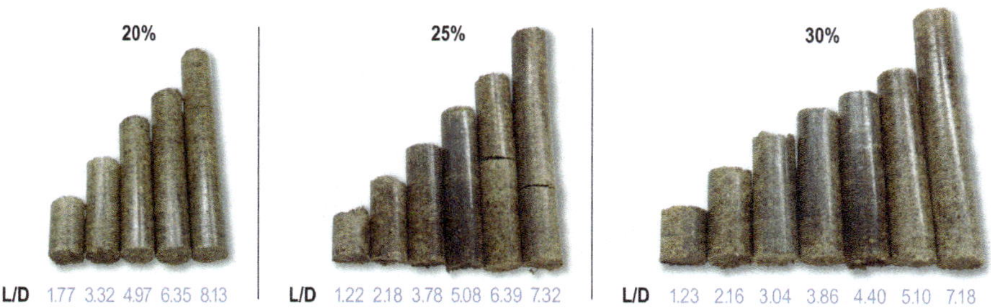

Fig. 3.6 Corn stover pellets made at different L/D ratios and feedstock moisture content values at 110 °C preheating temperature (derived from Tumuluru et al. 2017)

Fig. 3.7 Corn stover pellets made at different moisture contents, 110 °C preheating temperature, and 140 MPa (derived from Tumuluru et al. 2017)

Kawakita-Lüdde models show that the process is scalable; however, this must be validated in a pilot-scale pelleting system.

Single pelleting study results show that increasing the preheating temperature, decreasing the moisture content value, and increasing the L/D ratio results in a higher pellet unit density, matching with the data received from the laboratory-scale flat die pellet mill experiments (Tumuluru 2014, 2015, 2016; Tumuluru et al. 2017). Also,

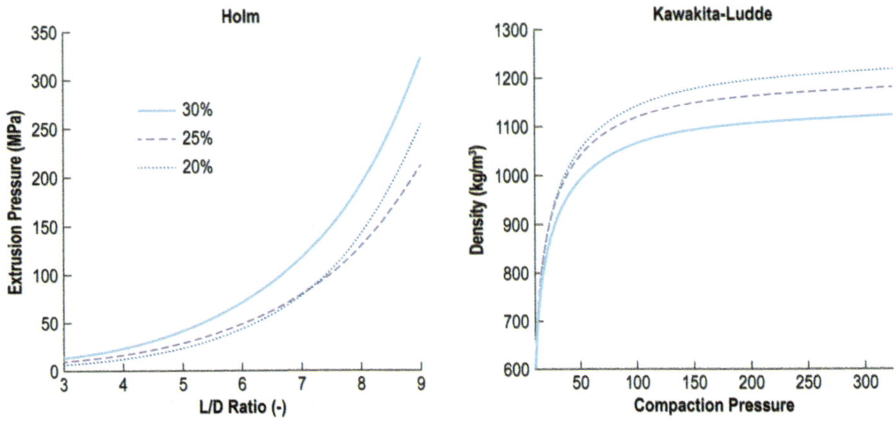

Fig. 3.8 Family of curves for corn stover at 110 °C for different L/D ratio and compression pressure based on Holm and Kawakita-Lüdde model (derived from Tumuluru et al. 2017)

single pelleting studies indicate that moisture content value is an important factor in influencing the density of a product, as observed in our earlier research on the flat die pellet mill. The unit densities of the pellets produced at 25% and 30% (w.b.) using an Instron matched closely with the laboratory-scale flat die pellet mill data, as shown previously in Fig. 3.8. Dynamic mechanical analysis (DMA) studies have validated the hypothesis that the glass transition temperature of lignin and other components is lower at higher moisture content values, which helps to bind biomass particles at higher moisture content values. The single pellet tests agree with flat die laboratory-scale tests in confirming the DMA results for lignin glass transition (Tg) temperatures for the high-moisture pelleting process.

3.1.3 Testing the Scale-Up Model in a Pilot-Scale Ring Die Pellet Mill

The scale-up model developed for the corn stover pelleting process using the Holm and Kawakita-Lüdde models was further tested in a pilot-scale ring die pellet mill. Based on the scale-up process at an L/D ratio, the extrusion pressure was about 80 MPa for a 20% (w.b.) moisture content value. At a steady-state pellet production process, the extrusion and compaction pressures balance each other out. At 200 MPa, the predicted value of the unit density of the produced corn stover pellet was about 1100 kg/m^3. Experiments were conducted to test the process in a pilot-scale ring die pellet mill that was fitted with a die having an L/D ratio of nine. Corn stover pelleting was conducted at different moisture content values ranging from 15 to 30% (w.b). The unit density of the produced pellets was about 1180 kg/m^3, which was close to the predicted value of 1150 kg/m^3.

3.1.4 *Response Surface Models and Optimization of the Pelleting Process*

3.1.4.1 Experimental Design

Experimental designs help in product development with a minimum number of experimental runs. The experiments are carried out for each variable at different levels. Some of the experimental designs commonly used in product design and process optimization are factorial, central composite, Box-Behnken, and rotatable. According to Mullen and Ennis (1979), for product development, the rotatable design is the best as it reduces the number of experiments for the process variables that are tested. In the experimental design, each variable has a different value range over which the variable will vary between upper and lower limits. Combining all the levels for the tested variables gives the total experimental combinations. Experimental design has been used by many researchers to develop response surface models (RSMs) for process optimization and product development (Shankar and Bandyopadhyay 2004, 2005; Shankar et al. 2010; Rout and Bandyopadhyay 1999; Bandyopadhyay and Rout 2001). These RSMs can be used for process optimization using statistical or evolutionary methods (Khuri and Cornell 1987; Frazier et al. 1983; Rout and Bandyopadhyay 1999; Giri and Bandyopadhyay 2000; Sacchetti et al. 2004; Shankar and Bandyopadhyay 2004, 2005).

According to Montgomery (1976), RSM is a collection of mathematical and statistical techniques used in the modeling and optimization of the process. The general form of the RSM is found in Eq. (3.14) (Montgomery 1976):

$$y = f(x_1, x_2, x_3, \ldots x_n) + \varepsilon \tag{3.14}$$

where:

y = response or dependent variable
$x_1, x_2, x_3, \ldots x_n$ = process or independent variables
ε = observed noise or error in measured the response variable y

With RSM, the first step is to find a suitable approximation for the true functional relationship between y and the independent variables tested. In general, a lower order polynomial in the independent variables tested is considered. If a linear model is well fitted for the independent variables, then the approximate model fitted is a first-order model, as shown in Eq. (3.15). If the process involves curvature, then a second-order polynomial model is used, as shown in Eq. (3.16). These models can help to identify the optimum process conditions for obtaining the desirable product quality attributes (Montgomery 1976):

$$y = \beta_0 + \beta_1 x_1 + \beta_2 x_2 + \beta_3 x_3 + \ldots \beta_k x_k + \varepsilon \tag{3.15}$$

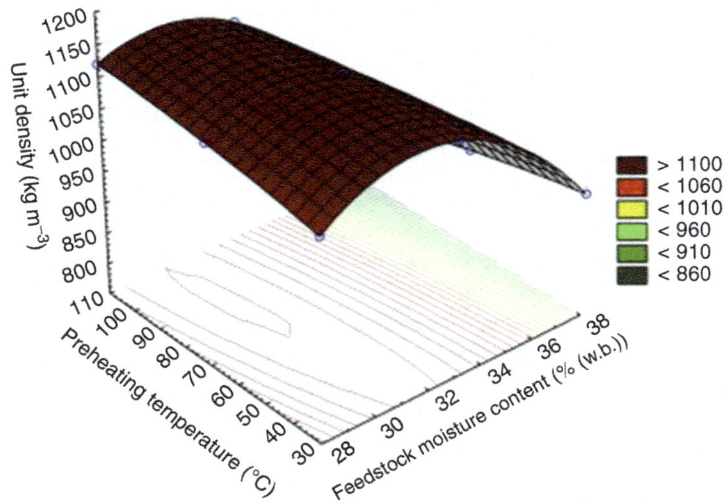

Fig. 3.9 The effect of preheating temperature and feedstock moisture content on unit density (derived from Tumuluru 2014)

$$y = \beta_0 + \sum_{i=1}^{k} \beta_i x_i + \sum_{i=1}^{k} \beta_{ii} {x_i}^2 + \sum \sum_{i<j} \beta_{ij} x_i x_j + \varepsilon \quad (3.16)$$

Tumuluru (2014, 2016) used RSM for the optimization of the pelleting of woody and herbaceous biomass to understand the interaction effect of the process variables on product properties. Tumuluru (2014, 2016) also used the Box-Behnken design to understand the impact of the process variables on pellet quality attributes, such as density and durability. This author developed second-order RSMs to describe the pelleting process with respect to the process variables, such as feedstock moisture content value, the rotational speed of the die, preheating temperature on pellet quality attributes (such as unit, and bulk density and durability), and specific energy consumption. Using these models, the author developed surface plots, as shown in Figs. 3.9, 3.10, and 3.11, respectively (Tumuluru 2014).

RSM is probably the most common optimization method and has been widely used for process modeling and optimization (Khan et al. 2005; Ravindra and Chattopadhyay 2000; Madamba and Lopez 2002; Azoubel and Murr 2004). In their research, Shankar et al. (2010) suggested the interpretation of the results becomes difficult if the process is very complex, primarily when the process variables are greater than three.

Biomass preprocessing processes—such as densification using an extruder, briquette press, or pellet mill, or grinding using a hammermill, rotary, or collision mill—are complex and involve both the systems and feedstock variables. These complex problems can be successfully optimized using evolutionary algorithms,

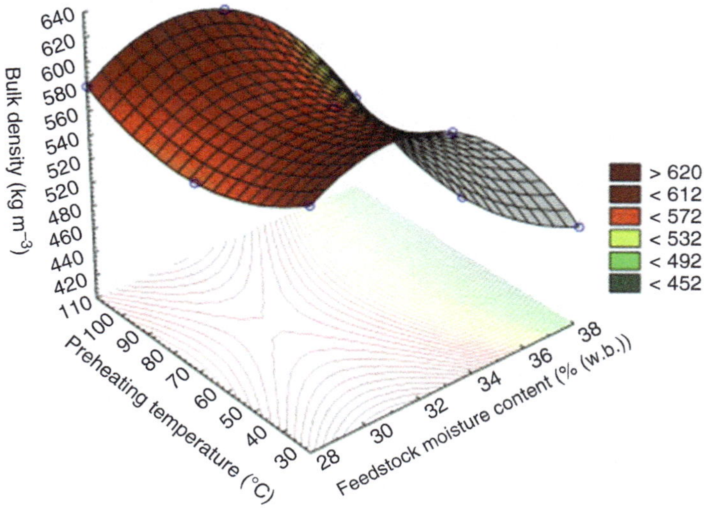

Fig. 3.10 The effect of preheating temperature and feedstock moisture content on bulk density (derived from Tumuluru 2014)

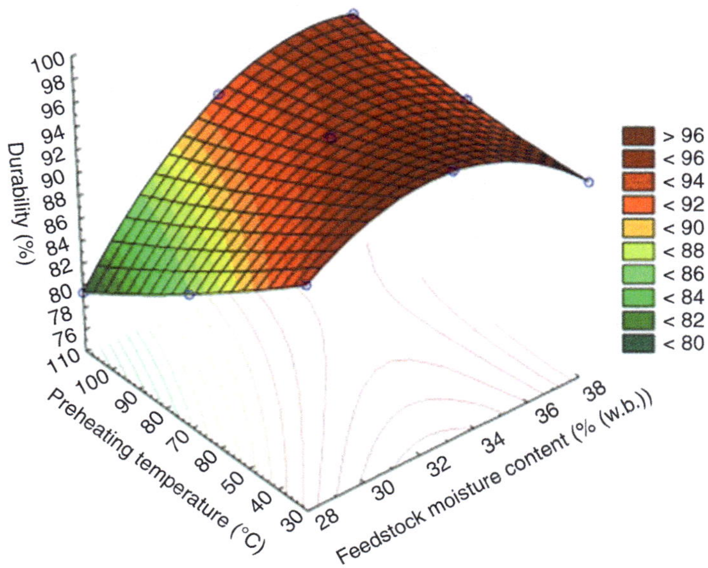

Fig. 3.11 The effect of preheating temperature and feedstock moisture content on durability (derived from Tumuluru 2014)

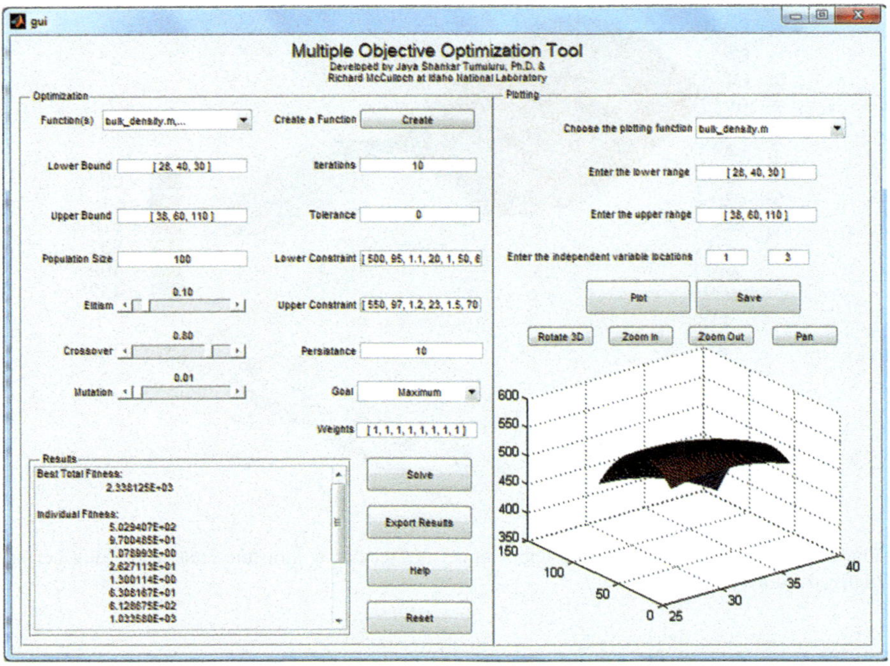

Fig. 3.12 Hybrid GA software, developed by Tumuluru and McCulloch (derived from Tumuluru and McCulloch 2016)

such as genetic algorithms (GAs). The evolutionary algorithm can solve a complex problem where deterministic methods might fail. Tumuluru and McCulloch (2016) suggested that stochastic algorithms—such as a GA—have shown more efficiency in optimizing these complex problems.

These authors developed a new hybrid GA and tested it for biomass unit operation grinding and pelleting optimization, as shown in Fig. 3.12. Shankar and Bandyopadhyay (2004, 2007) and Shankar et al. (2010) successfully used GAs and artificial neural networks to understand and optimize real-life time problems, such as extrusion processes. These authors used a combination of RSM and GA to understand better and optimize the pelleting process. Their studies were aimed at figuring out how process variables (such as barrel temperature, screw speed, and L/D ratio) and feedstock variables (such as feed moisture content and feed content of the biomass mix) affect flow behavior and quality attributes (such as expansion ratio, bulk density, and hardness). Shankar and Bandyopadhyay (2004), Tumuluru and McCulloch (2016), Tumuluru (2019), and Tumuluru and Heikkila (2019) tested the genetic and hybrid GAs for the optimization of food and biomass processing (such as the grinding and pelleting process). They found that the RSMs developed for more than three variables can be successfully optimized using evolutionary algorithms (such as genetic and hybrid GAs). There are numerous optimization algorithms—

such as the ant colony (ACO) algorithm, simulated annealing (SA), tabu search, and particle swarm optimization (PSO)—which are widely used for the optimization of complex problems. These algorithms work extremely well in finding real-time solutions to complex systems. The ACO is a probabilistic technique to solve computational problems. SA is a global optimization technique, which finds solutions by traversing the search space by generating neighboring solutions to the current solution. A tabu search functions in a similar fashion as SA, which tests the mutations of the individual solutions in the search space. The major difference between a SA and a tabu search is that the former generates only one mutated solution, whereas the latter generates many mutated solutions and moves the solution toward the lowest fitness values generated (Holland 1992; Deb 2000). On the other hand, the PSO algorithm belongs to direct search methods that can be used to find optimized solutions in a given search space for a given objective function. PSO algorithm has many advantages, such as the basic algorithm that can be implemented by the programmer quickly. Also, it can be used by a non-programmer for optimization of objective functions without any background to complex mathematical optimization theory.

3.2 Conclusions

In this chapter, the existing compression models were reviewed. The suitability of these equations for different biomasses was also presented. For agricultural biomass, the Kawakita-Lüdde model provided an excellent fit having R^2 values of 0.99. The parameter of the model provides a good parameter to understand compression characteristics. For example, studies conducted by the researchers indicate the parameters of the Cooper-Eaton model show the ground straw samples were densified easily by the particle rearrangement method. In contrast, the Jones model indicated that canola and oat straw was more compressible as compared to the barley and wheat straw. In addition, the compression models can be used for the scale-up of the pelleting process. Data obtained using the Holm and Kawakita-Lüdde models can be used further to develop the relationship for compression pressure to density and L/D ratio to extrusion pressure, which can, in turn, be used to identify the desired L/D ratio necessary to produce pellets with desired density. New algorithms, such as genetic and hybrid GAs, can be used in combination with RSM to optimize densification process models.

References

Adapa PK, Bucko J, Tabil Jr. LG, Schoenau G, Sokhansanj S (2002) Pelleting characteristics of fractionated sun-cured and dehydrated alfalfa grinds. ASAE/CSAE North-Central Intersectional Meeting, Saskatoon, Saskatchewan, Canada. 27–28 September 2002

Adapa PK, Tabil Jr. LG, Schoenau G (2009) Compression characteristics of selected ground agricultural biomass. Ag Eng Int: CIGR Ejournal Manuscript 1347, vol XI. June 2009

Alderborn G, Wikberg M (1996) Granule properties. In: Alderborn G, Nyström C (eds) Pharmaceutical Powder Compaction Technology. Marcel Dekker, New York, pp 323–373

Azoubel P, Murr F (2004) Mass transfer kinetics of osmotic dehydration of cherry tomato. J Food Eng 61:291–295

Bandyopadhyay S, Rout RK (2001) Aquafeed extrudate flow rate and pellet characteristics from low-cost single-screw extruder. J Aquat Food Prod T 10(2):3–15

Comoglu T (2007) An overview of compaction equations. J Fac Pharm Ankara 36(2):123–133

Cooper AR, Eaton LE (1962) Compaction behavior of several ceramic powders. J Am Ceram Soc 45(3):97–101

Deb S (2000) Optimization for engineering design: algorithms and examples. Prentice-Hall of India Private Ltd, New Delhi

Denny PJ (2002) Compaction equations: A comparison of the Heckel and Kawakita equations. Powder Technol 127:162–172

Frazier PH, Crawshaw A, Daniels NWR, Eggitt PWR (1983) Optimization of process variables in extrusion cooking of soya. In: Jowitt R (ed) Extrusion Cooking Technology. Elsevier Applied Science Publishers, London, England, pp 1–26

Giri SK, Bandyopadhyay S (2000) Effect of extrusion variables on extrudate characteristics of fish muscle-rice flour blend in single screw extruder. J Food Process Press 24:177–190

Gray WA (1968) Compaction after deposition. In: Gray WA (ed) The Packing of Solid Particles. Chapman and Hall, New York, pp 89–107

Heckel RW (1961) An analysis of powder compaction phenomena. T Metall Soc AIME 221:1001–1008

Holland JH (1992) Genetic algorithms. Sci Am 267:66–72

Holm JK, Henriksen UB, Hustad JE, Sorensen LH (2006) Toward an understanding of controlling parameters in softwood and hardwood pellets production. Energy Fuel 20:2686–2694

Johansson B, Alderborn G (1996) Degree of pellet deformation during compaction and its relationship to the tensile strength of tablets formed of microcrystalline cellulose pellets. Int J Pharm 132(1–2):207–220

Johansson B, Wikberg M, Ek R, Alderborn G (1995) Compression behaviour and compactability of microcrystalline cellulose pellets in relationship to their pore structure and mechanical properties. Int J Pharm 117(1–2):57–73

Jones WD (1960) Fundamental principles of powder metallurgy. Edward Arnold Publishers Ltd, London, England, pp 242–370

Kawakita K, Lüdde KH (1971) Some considerations on powder compression equations. Powder Technol 4(2):61–68

Khan M, Andres A, Shankar T, Oliveira F, Cuhna L (2005) Analysis of the effect of process variables on the osmotic dehydration of mango and process optimization using a genetic algorithm approach. In: Proceedings of the III International Symposium on Applications of Modelling as an Innovative Technology in the Agri-Food Chain, Leuven, Belgium. 29 May 2005, 674:181–188

Khuri AZ, Cornell JA (1987) Response surfaces: Designs and Analyses. Marcel Dekker Inc., New York, NY, USA

Madamba PS, Lopez RI (2002) Optimization of the osmotic dehydration of mango (*Mangifera Indica* L.) slices. Dry Technol 20(6):1227–1242

Mani S, Tabil LG Jr, Sokhansanj S (2003) An overview of compaction of biomass grinds. Powder Hand. Process 15(3):160–168

Mani S, Tabil LG Jr, Sokhansanj S (2004) Evaluation of compaction equations applied to four biomass species. Can Biosyst Eng 46:3.55–3.61
Montgomery DC (1976) Design and Analysis of Experiments, 4th edn. Wiley, New York, NY, USA
Mullen K, Ennis DM (1979) Rotatable design in product development. J Food Technol 33 (7):74–75, 78–80
Panelli R, Filho FA (2001) A study of a new phenomenological compacting equation. Powder Technol 114:255–261
Ravindra M, Chattopadhyay P (2000) Optimization of osmotic preconcentration and fluidized bed drying to produce dehydrated quick-cooking potato cubes. J Food Eng 44:5–11
Rout RK, Bandyopadhyay S (1999) A comparative study of shrimp feed pellets processed through cooking extruder and meat mincer. Aquac Eng 19:71–79
Rumpf H (1962) The strength of granules and agglomerates. In: Knepper WA (ed) Agglomeration. Interscience Publishers, New York, NY, USA, pp 379–418
Sacchetti G, Pinnavaia GG, Guidolin E, Dalla Rosa M (2004) Effects of extrusion temperature and feed composition on the functional, physical, and sensory properties of chestnut and rice flour-based snack-like products. Food Res Int 37(5):527–534
Sastry KVS, Fuerstenau DW (1973) Mechanisms of agglomerate growth in green pelletization. Powder Technol 7:97–105
Shankar TJ, Bandyopadhyay S (2004) Optimization of extrusion process variables using a genetic algorithm. Food Bioprod Process 82(2):143–150
Shankar TJ, Bandyopadhyay S (2005) Process variables during single screw extrusion of fish and rice flour blends. J Food Process Preserv 29:151–164
Shankar TJ, Bandyopadhyay S (2007) Prediction of extrudate properties using artificial neural networks. Food Bioprod Process 85(1):29–33
Shankar TJ, Sokhansanj S, Bandyopadhyay S, Bawa AS (2010) A case study on optimization of biomass flow during single-screw extrusion cooking using genetic algorithm (GA) and response surface method (RSM). Food Bioprocess Tech 3:498–510
Shivanand P, Sprockel OL (1992) Compaction behavior of cellulose polymers. Powder Technol 69:177–184
Sonnergaard JM (2001) Investigation of a new mathematical model for compression of pharmaceutical powders. Eur J Pharm Sci 14:149–157
Stelte W, Sanadi A, Shang L, Holm J, Ahrenfeldt J, Henriksen U (2012) Recent developments in biomass pelletization—A review. Bioresources 7(3):4451–4490
Tabil LG Jr, Sokhansanj S (1996) Process conditions affecting the physical quality of alfalfa pellets. Appl Eng Agric 12(3):345–350
Tumuluru JS (2014) Effect of process variables on the density and durability of the pellets made from high moisture corn stover. Biosyst Eng 119:44–57
Tumuluru JS (2015) High moisture corn stover pelleting in a flat die pellet mill fitted with a 6 mm die: physical properties and specific energy consumption. Energy Sci Eng 3(4):327–341
Tumuluru JS (2016) Specific energy consumption and quality of wood pellets produced using high moisture lodgepole pine grind in a flat die pellet mill. Chem Eng Res Des 110:82–97
Tumuluru JS (2018) Effect of pellet die diameter on density and durability of pellets made from high moisture woody and herbaceous biomass. Carbon Resour Convers 1(1):44–54
Tumuluru JS (2019) Pelleting of pine and switchgrass blends: effect of process variables and blend ratio on the pellet quality and energy consumption. Energies 12(7):1198
Tumuluru JS, Heikkila DJ (2019) Biomass grinding process optimization using response surface methodology and a hybrid genetic algorithm. Bioengineering 6(1):12
Tumuluru JS, McCulloch R (2016) Application of hybrid genetic algorithm routine in optimizing food and bioengineering processes. Foods 5(4):76
Tumuluru JS, Yancey N, McCulloch R, Fox C, Conner CC, Hartley D, Dee M, Plummer M (2017) Biomass engineering: size reduction, drying, and densification of high moisture biomass. Feedstock supply and logistics platform, 2017 project peer review, US Department of Energy (DOE) Bioenergy Technologies Office (BETO), 7 March 2017, Denver, CO, USA. https://www.energy.gov/sites/prod/files/2017/05/f34/fsl_tumuluru_1222.pdf. Accessed 8 Apr 2020

van Dam JEG, van den Oever MJA, Teunissen W, Keijsers ERP, Peralta AG (2004) Process for production of high density/high performance binderless boards from whole coconut husk—Part 1: Lignin as intrinsic thermosetting binder resin. Ind Crop Prod 19(3):207–216

Walker EE (1923) The properties of powders—Part VI: The compressibility of powders. T Faraday Soc 19:73–82

York P, Pilpel N (1972) The effect of temperature on the mechanical properties of some pharmaceutical powders in relation to tableting. J Pharm Pharmacol 24:47P–56P

Chapter 4
Advances in Biomass Densification and Its Impact on Cost and Quality

Abstract Even though densified products have many advantages in terms of storage, handling, transportation, and conversion performance, existing methods to produce dense biomass pellets are currently not considered due to the prohibitive costs to densify them. At present, new preprocessing technologies are developed to address the cost and quality issues associated with conventional methods. This chapter will focus on the advances achieved by industry and national laboratories in terms of biomass size reduction and densification that have a significant impact on energy consumption, cost, and quality. Also, the unit operations and their impact on conventional and advanced preprocessing unit operations are discussed in this chapter.

Keywords Rotary shear grinding · Fractional milling · High-moisture pelleting · Low-temperature drying · Dewatering · Techno-economic analysis

4.1 Introduction

The feedstock handling systems at biorefineries have historically experienced problems typically resulting from variability in the physical characteristics of the material, including composition, moisture content, particle size, and throughput, thereby causing inconsistent conversion of mass and heat transfer within the thermal conversion process. Herbaceous biomass used for bioenergy applications is more homogenous in terms of their physical properties and chemical compositions. Feedstocks like municipal solid waste (MSW) are more heterogeneous. They contain a wide variety of materials (e.g., wood, paper, plastic, and other organics) that comprise the waste stream, resulting in less consistent feed handling behavior as compared to homogenous feedstocks.

A 1980s-era Rand Corporation study showed that plants processing bulk solids typically operate at less than 50% of design capacity during their first year of operation, the primary reason being an inadequate understanding of the behavior of particle systems. As depicted in Fig. 4.1, feedstock type can impact system efficiency and operability by causing plugging of bins, augurs, downstream

© Springer Nature Switzerland AG 2020

J. S. Tumuluru, *Biomass Densification*,
https://doi.org/10.1007/978-3-030-62888-8_4

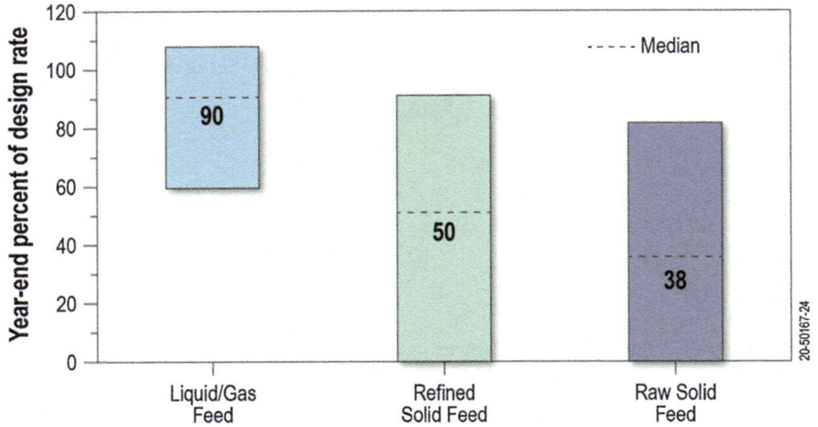

Fig. 4.1 Feedstock type impact on the operability of the refineries

equipment, and microbial contamination due to the non-uniform cooking of particles in the reactor (Bell 2005).

In general, particles can almost be described as the fourth state of matter. They can develop cohesive strength and transfer stresses like a solid. They can also retain air, take on fluid-like properties, and may be compressible and elastic. Gases and liquids do not grow, but the particles agglomerate, aggregate, or suffer attrition, and their behavior often does not scale.

4.1.1 Techniques to Overcome the Particle Handling Problems

There is a wide range of alternative techniques to address these particle handling problems. They include design improvements, control systems, and material engineering. Some of the techniques that are commonly used to overcome the particle systems limitations are described in more detail in the sections that follow.

4.1.1.1 Design Improvements

The first technique that can be used to address these issues is to implement design improvements in the preprocessing system and/or handling equipment based on the biomass material properties. However, this technique can be limited if: (a) a system is being retrofitted; (b) the wrong symptom is being targeted, thereby only leading to the creation of another challenge; (c) based on material properties (e.g., flow behavior) that are empirical and subjective; and (d) designed only to the worst-case scenario, which could be cost-prohibitive.

4.1.1.2 Control Systems

The second technique is to develop control systems that monitor and adjust processing conditions to maintain a consistent feedstock quality and production rate. Examples include reactive controls, which keep a more constant mass flow, and adaptive control that maintains feedstock specifications. This technique requires: (a) a good understanding of the infeed material properties, particle characteristics, and particle interactions with process equipment, which can be unpredictable; and (b) the installation of added controls (e.g., software, sensors, and actuators) to create a fully integrated system. The benefits include maximizing processing system and equipment performance and control data, which provide additional knowledge of particle systems and the opportunity to improve equipment and material engineering solutions further.

4.1.1.3 Material Engineering

The third technique is to implement material engineering solutions. With this technique, new systems would be inserted into existing processing systems that are designed to alter biomass material properties. This includes: (a) densification of biomass to overcome compressible and elastic behavior; (b) the addition of flow additives to address cohesiveness; and (c) heat treatment to cause the mild deconstruction of the biomass cell structures. The benefits of this technique include implementing a solution designed to address a specific material property and scalability of the solution. Limitations may include an increase in capital and operational costs and adaptability to changing material properties. Further analysis and assessment must be conducted to better understand the feasibility, efficiency, and costs for engineering solutions and whether the capital costs are outweighed in terms of value added by improvements in feed handling, conveyance, and storage safety.

4.1.1.4 Conventional Densification Process

Variability in the physical properties of the biomass—such as density, moisture, particle size, shape, and composition—makes the techniques used for processing, handling, and transporting the biomass-derived feedstocks difficult. Variability in moisture will not only create complexities with feed handling, but could also increase costs associated with additional preprocessing and drying equipment.

Biomass preprocessing is done to make the product aerobically stable, as well as easy to transport, flow, and feed through the conversion process to produce biofuels. The general preprocessing steps for woody and herbaceous biomass include size reduction, drying, and densification. Size reduction includes unit operations—such as chipping, grinding, and chopping—using different grinders—such as hammermills, knife mills, collision mills, and shear mills. Drying is done to make

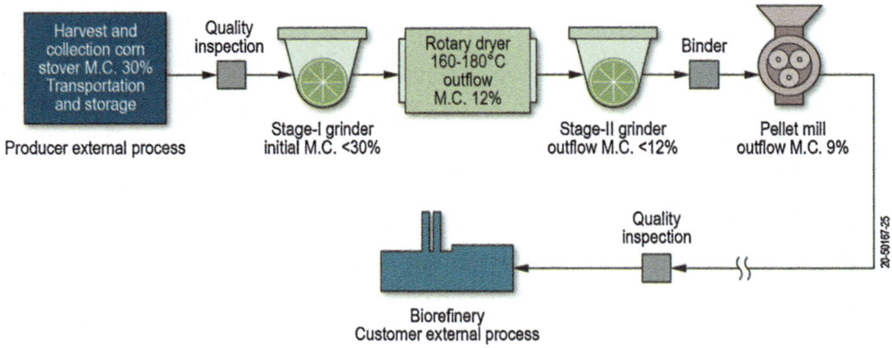

Fig. 4.2 Various unit operation in the conventional pelleting process (derived from Lamers et al. 2015)

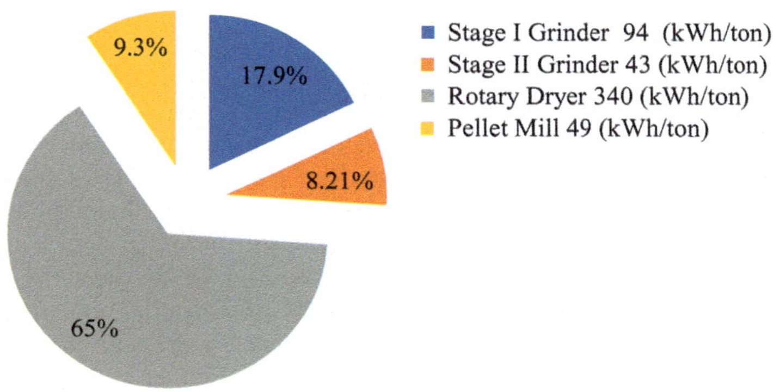

Fig. 4.3 Energy consumption for the different unit operations in lodgepole pine biomass pelleting (derived from Tumuluru 2015)

the biomass aerobically stable. Typically, both woody and herbaceous MSW biomass is available at moisture content values of >30% (w.b.). At higher moisture contents, biomass loses dry matter during storage, which results in the lowering of its energy content. Drying or removing the moisture from the biomass is critical to keep it aerobically stable. Typically, biomass is dried to <10% (w.b.) using high-temperature dryers, such as rotary dryers. Biomass drying also reduces its size. This dried biomass can then be further densified in a pellet mill, briquette press, or cuber to improve its physical properties, such as density, as well. Figure 4.2 shows the various stages and energy consumption requirements typically used in the pellet production process.

Conventional pellet production processes used in biomass energy facilities can be capital- and energy-intensive, as seen in Fig. 4.2 (Tumuluru 2015). Figure 4.3 indicates the energy consumption of various unit operations in pellet production processes. Table 4.1 shows the energy consumption of pellet production from woody and herbaceous biomass and their blends. Drying using rotary dryers takes about

Table 4.1 Energy constipation of various unit operations in the pellet production of woody and herbaceous biomass pellets (derived from Yancey et al. 2013)

	Conventional pelleting unit operations			
Biomass type	Stage I grinding (kWh/ton)	Rotary drying (kWh/ton)	Stage II grinding (kWh/ton)	Pelletization (kWh/ton)
4 Crop (corn stover+switchgrass +eucalyptus+lodgepole pine) mix	54.83	185.64	56.52	76.36
Switchgrass	11.72		18.55	45.20
Eucalyptus	94.11	401.78	49.97	124.13
Lodgepole pine	94.11	340.77	43.21	48.94
Corn stover	19.38		34.54	56.43

65–70% of the total energy of the process. Rotary dryers have several disadvantages, including the need for: (a) high-quality heat; (b) emission control equipment for particulate matter and volatile organic compound (VOC) emissions; (c) additional fire protection and suppression equipment; and (d) more accurate control of material moisture content (Tumuluru 2014). These emissions represent a significant environmental challenge in the high-temperature biomass drying process. Studies conducted by Bengtsson (2004) indicated that pine contains more extractives than spruce, and the hydrocarbon emissions depend on the initial concentration of the hydrocarbons in the biomass material. Studies by Rupar and Sanati (2003) on the drying of woody biomass have indicated that monoterpene and sesquiterpene hydrocarbons, and a negligible amount of diterpene hydrocarbons, are emitted. Jukka-Pekka and Pekka (2003) reviewed the organic compounds emissions during the drying of biomass. At temperatures below 200 °C, most of the compounds that are emitted are lipophilic compounds, where at temperatures higher than 200 °C, hydrophilic extractives and wood particulate compounds are observed. The emissions are about 2–5% (d.b.) for coniferous compounds and 2–4% (d.b.) for those coming from broad-leaved trees (Jukka-Pekka and Pekka 2003). The VOC emissions from the drying of hardwoods and softwoods are methanol and formaldehyde—both hazardous environmental pollutants (Jukka-Pekka and Pekka 2003). Also, these same authors indicated that low-temperature drying helps to reduce VOC emissions significantly. In terms of investment, rotary driers, pelletizers, and belt dryers all require additional capital.

Figure 4.4 indicates the changes in the bulk density of baled and densified biomass. Densification increases bulk density by three to five times. Figure 4.5 shows the durability of the pellets produced using various woody and herbaceous biomass feedstocks and their blends. One way to reduce the cost of pellet production is to manage moisture more efficiently. The major challenge is the cost of pellet production. According to Lamers et al. (2015), it takes about $65 to produce 1 ton wood pellets, as observed in Fig. 4.6.

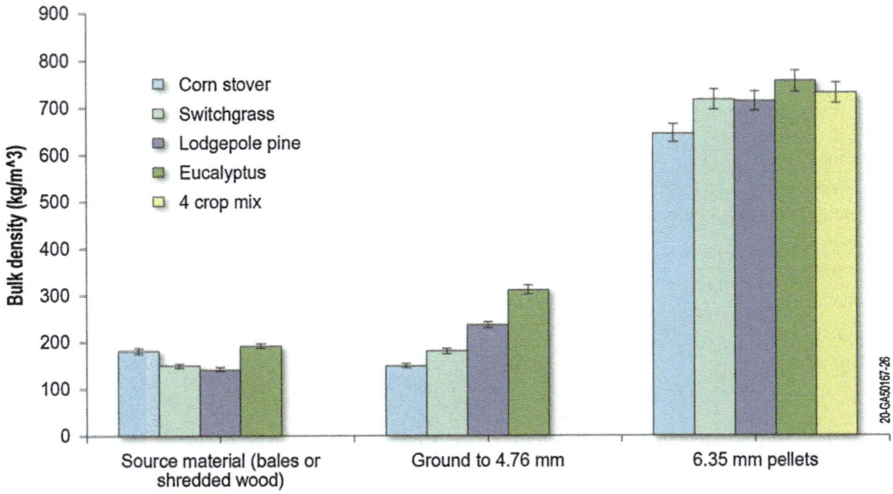

Fig. 4.4 Bulk density of raw, ground, and pelleted woody and herbaceous material (derived from Yancey et al. 2013)

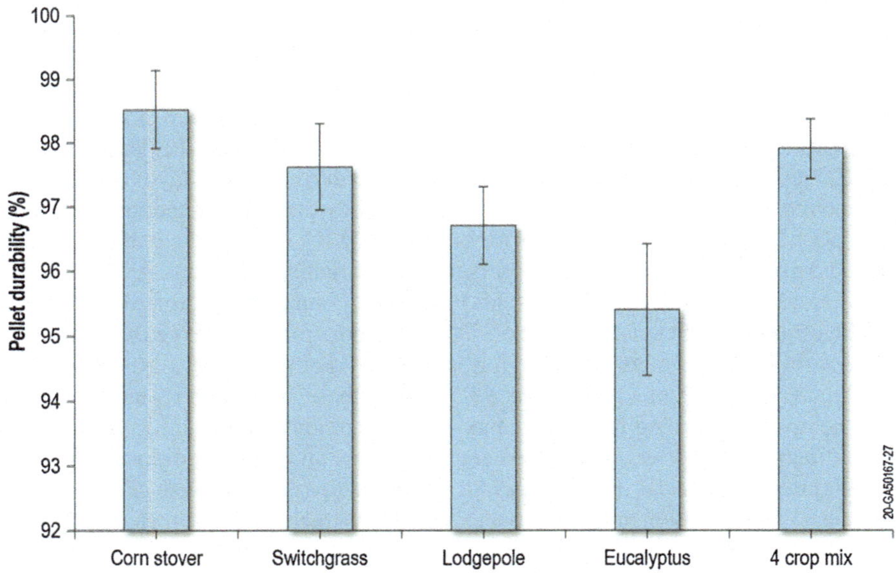

Fig. 4.5 The durability of herbaceous, woody, and formulated biomass pellets (derived from Yancey et al. 2013)

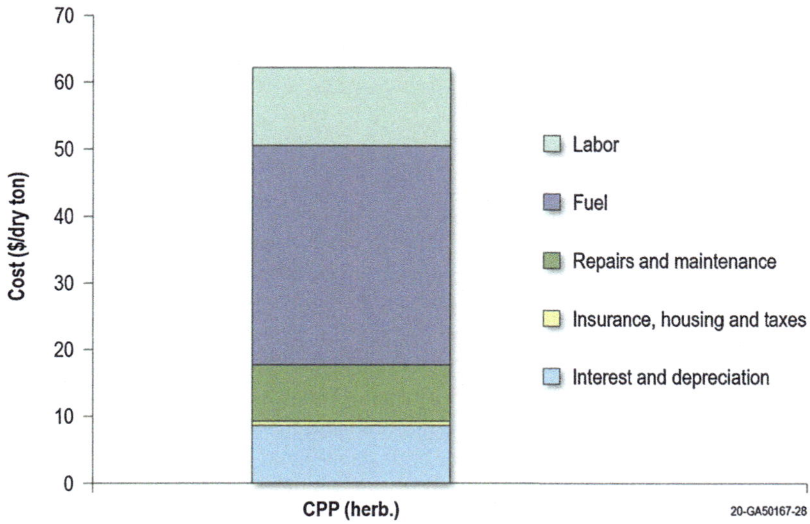

Fig. 4.6 Techno-economic analysis of the conventional pelleting process (derived from Lamers et al. 2015)

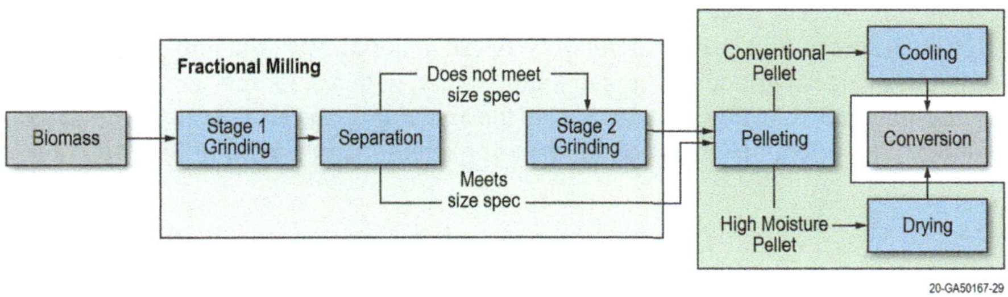

Fig. 4.7 Cost-effective techniques for moisture management (derived from Tumuluru and Yancey 2018)

4.1.1.5 Efficient Moisture Management Through Preprocessing

Idaho National Laboratory (INL) has developed advanced preprocessing technologies that can help to manage moisture more efficiently, thereby reducing preprocessing costs (Fig. 4.7). Kenney et al. (2013) presented a feasible approach to reduce preprocessing costs. According to these authors, a combination of technologies such as fractional milling to reduce grinding costs, high-moisture pelleting, and low-temperature can reduce preprocessing costs by more than 50% as compared to 2013 state of technology (SOT) for corn stover pellet production, as observed in Fig. 4.6. The final product delivered with 2013 SOT is a 1-inch (50.4 mm) grind; however, using the advanced preprocessing methods shown in Fig. 4.7 not only helps to manage moisture efficiently but also produces a pellet as the final product.

Converting biomass into pellets helps to address both moisture and particle size variability, as well as elastic and cohesive property issues.

4.1.1.6 Fractional Milling

In the fractional milling process, the screen size of a stage-1 grinder is increased, and a separator is inserted between the stage-1 and stage-2 grinders to bypass the material or fraction that has already met the specifications (Kenney et al. 2013; Tumuluru and Yancey 2018). This process helps to decouple stage-1 and stage-2 grinding, save energy, and avoid redundant processing and tighter particle size distribution with reduced fines. Figures 4.8 and 4.9 indicate the energy consumption for the stage-1 and stage-2 corn stover grinding processes. It is very clear from the figures that the moisture content in the biomass and the screen size of the grinder are inversely related to grinding energy for both stage-1 and stage-2 grinders and that the grinding energy increases exponentially with smaller screen sizes in both of the stage-1 and stage-2 grinders at higher moisture content values of 30% (w.b.). One way to reduce the grinding energy of high-moisture biomass is to introduce a separator between the stage-1 and stage-2 grinders and bypass the material that has already met the required specifications. Figure 4.10 shows the separator efficiency for different screen sizes. It is clear from Fig. 4.10 that a separator fitted with a 7/16-inch screen will bybass about 50% of the material, with the remaining material sent to the stage-2 grinder. If the separator is fitted with a 6.35 mm screen, about 35% of the material will be bypassed, with the remaining material sent to the stage-2 grinder. The effect of separation on grinding energy is provided in Fig. 4.11. From

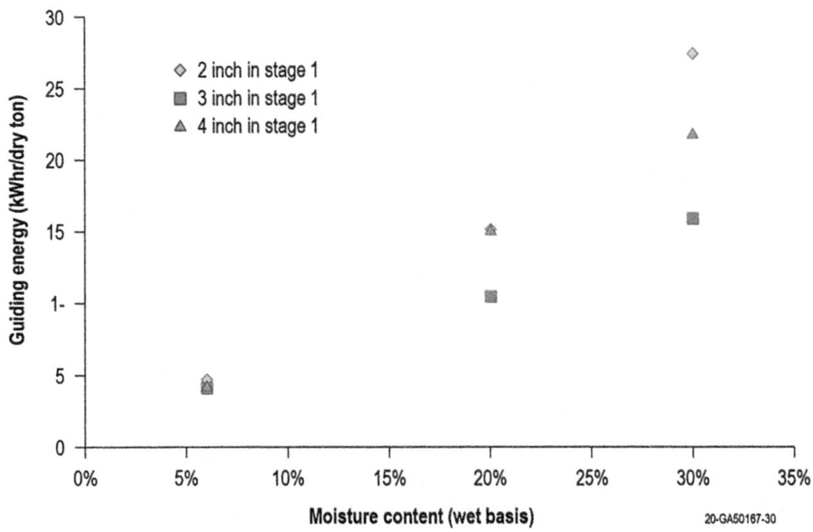

Fig. 4.8 The effect of moisture content and screen size on stage-1 grinding energy

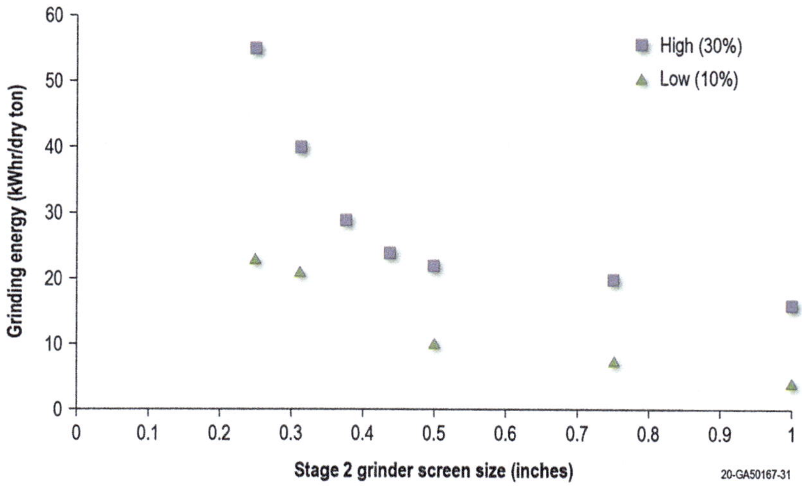

Fig. 4.9 The effect of moisture content and screen size on stage-2 grinding energy

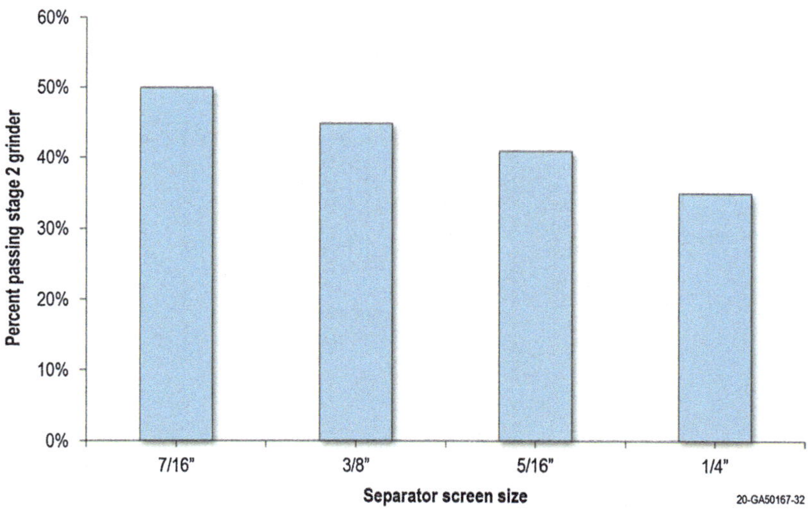

Fig. 4.10 The effect of separator screen size on separation efficiency

the figure, it is obvious that grinding energy can be significantly reduced by bypassing the material that has already met the required specifications.

Fractional milling studies were conducted to understand the impact of stage-1 grinder screen size, corn stover bale moisture content, and grinder feed rate. Figure 4.12 indicates the fractional milling process tested. Experiments were conducted on corn stover using fractional milling technology using the hammermills currently part of the Process Demonstration Unit (PDU) at INL. Data on separation efficiency

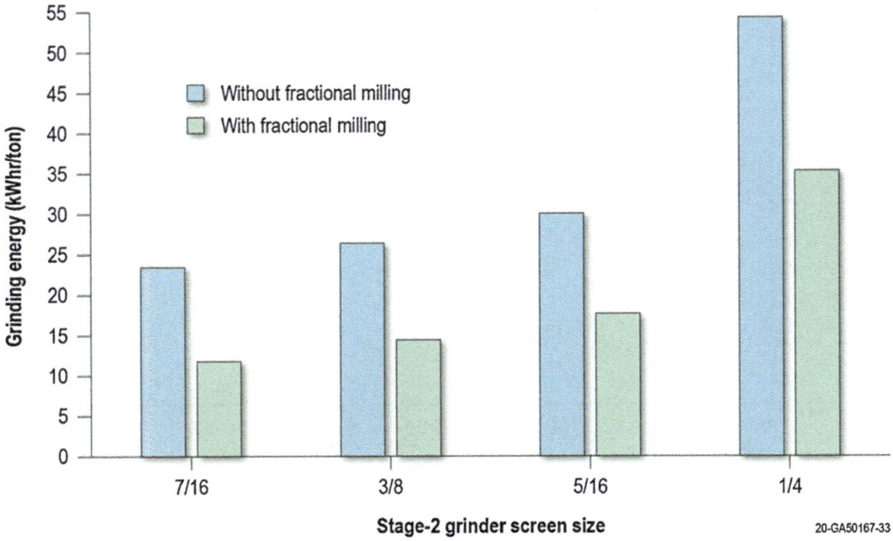

Fig. 4.11 The effect of separation efficiency on grinding energy

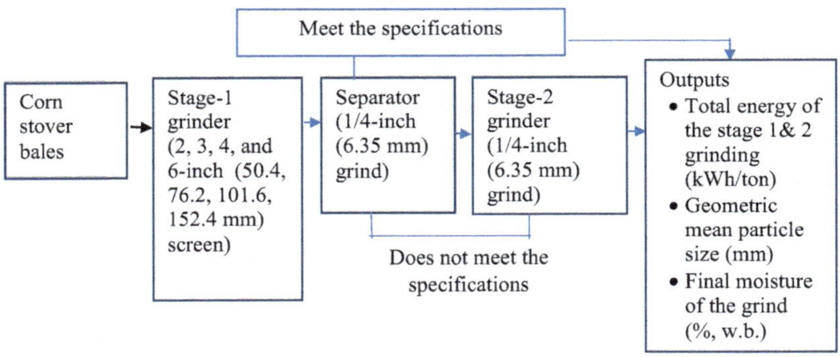

Fig. 4.12 Flow diagram of the fractional milling process tested for different moisture corn stover bales

Table 4.2 Process variables tested for corn stover fractional milling studies

Process variables	Lower limit	Upper limit
Stage-1 grinder screen size (inch) (x_1)	2-inch (50.4 mm)	6-inch (152.4 mm)
Bale moisture content (%, w.b.) (x_2)	5	25
Feed rate (%) (x_3)	2	12

were collected from testing at INL, Rotex Screeners, and Bulk Handling Systems BHS LLC. The process data on stage-1, stage-2, and separations were collected and modeled. Table 4.2 indicates the process variables used for the fractional milling studies. The product properties are geometric mean particle size, % bypassed, and final moisture of the grind. Also, the total grinding energy was measured.

Surface plots were developed to understand the effect of process variables on the product properties and total energy consumption. Surface plots were drawn for the grinding process variables (i.e., screen size, bale moisture content, and feed rate) with respect to total grinding energy and ground corn stover properties, such as geometric mean particle size, % bypassed, and final moisture content of the grind.

4.1.1.7 Total Grinding Energy (Stage-1 + Stage-2 Grinders)

Figures 4.13 and 4.14 show how total grinding energy (stage-1 + stage-2) changes with respect to bale moisture content, screen size, and feed rate. It is very clear from Fig. 4.13 that increasing bale moisture content had a greater impact on increasing the total grinding energy than increasing the screen size. Figure 4.14, which is drawn between feed rate and screen size, indicated that higher feeding rates of 12% and a bigger screen size of 6 mm help to reduce the total grinding energy to about <27 kWh/ton for bales at 20% moisture content. The same figure indicates that screen size has a less significant effect on the grinding energy compared to feed rate. Lowering the biomass feed rate increased the total grinding energy to >55 kWh/ton.

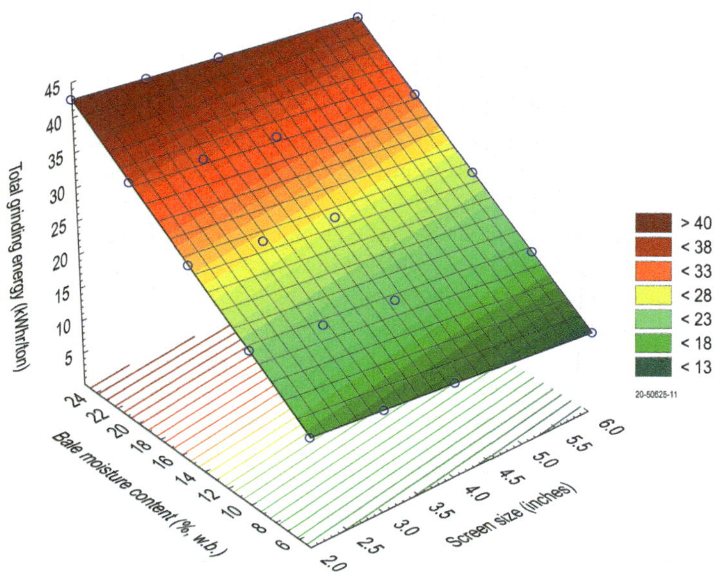

Fig. 4.13 Effect of stage-1 grinder screen size and bale moisture content on the total grinding energy at a 10% feed rate

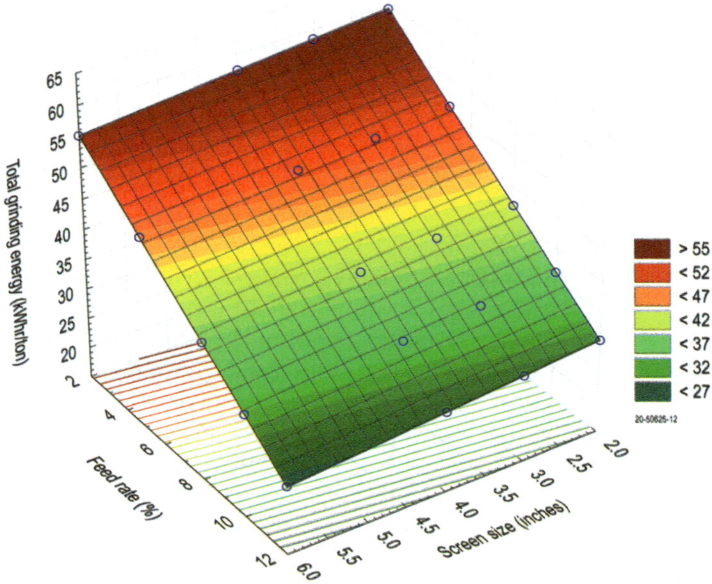

Fig. 4.14 Effect of feed rate and stage-1 grinder screen size on the total grinding energy at 20% bale moisture content

4.1.1.8 Geometric Mean Particle Size

Surface plots were also drawn for the geometric mean particle size of the particles produced during fractional milling. Figure 4.15 is drawn for the screen size of the stage-1 grinder and bale moisture content. The results indicated that both the screen size of the stage-1 grinder and bale moisture content has a significant effect on the geometric mean particle size. The geometric mean particle size increased with increasing bale moisture content and screen size. Using a 6-inch screen in the stage-1 grinder and a 1/4-inch screen in the stage-2 grinder at 25% bale moisture content produced a particle with geometric particle size >1.2 mm, whereas using a 2-inch screen in the stage-1 grinder and a 1/4-inch screen in the stage-2 grinder at a bale moisture content of 5% produced a geometric mean particle size <0.825 mm. For a 3-inch screen size in the stage-1 grinder, increasing the bale moisture content and feed rate increased the geometric mean particle size (Fig. 4.16). A maximum geometric mean particle size of >1.1 was observed at 25% (w.b.) bale moisture content and 12% feed rate. Figure 4.17 indicates that both bale moisture content and feed rate has a significant effect on the geometric mean particle size. The smallest mean particle size of <0.76 mm was observed at a bale moisture content of 5% and a feed rate of 2%. Figure 4.17, which is drawn for feed rate and screen size, indicates that lowering the feed rate to 2% and going for a smaller screen size of 2 inches in the stage-1 grinder at 20% bale moisture content reduces the geometric mean particle size to <0.9 mm.

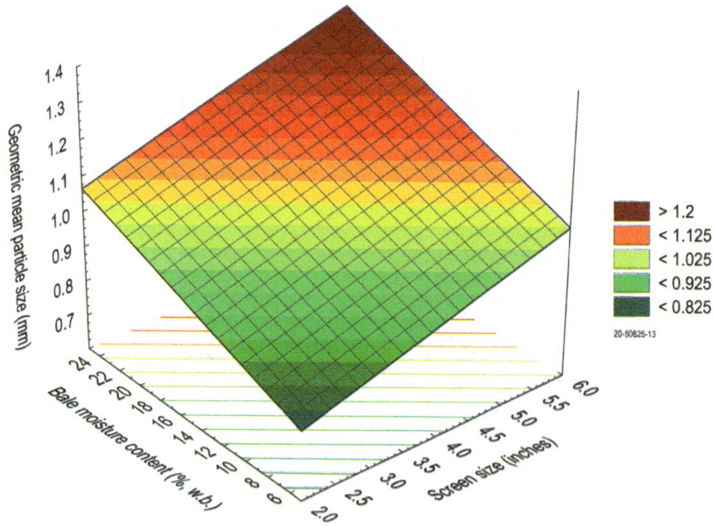

Fig. 4.15 Effect of stage-1 grinder screen size and bale moisture content on the geometric mean particle size at a 10% feed rate

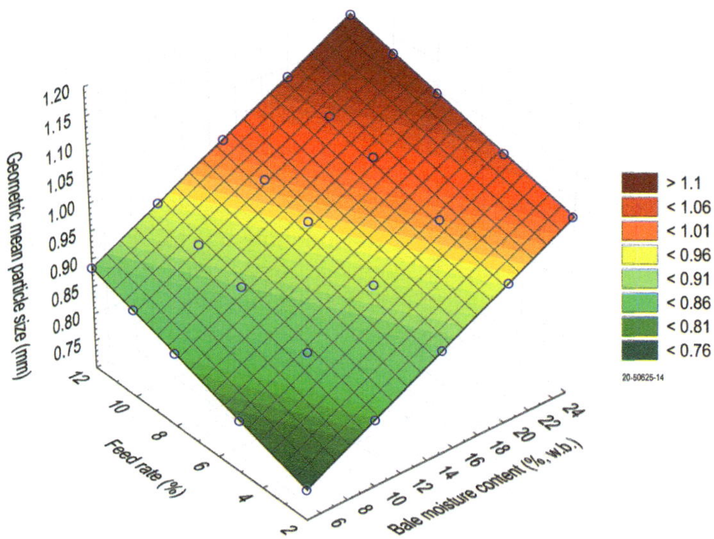

Fig. 4.16 Effect of bale moisture content and feed rate on the geometric mean particle size for 3-inch screen size

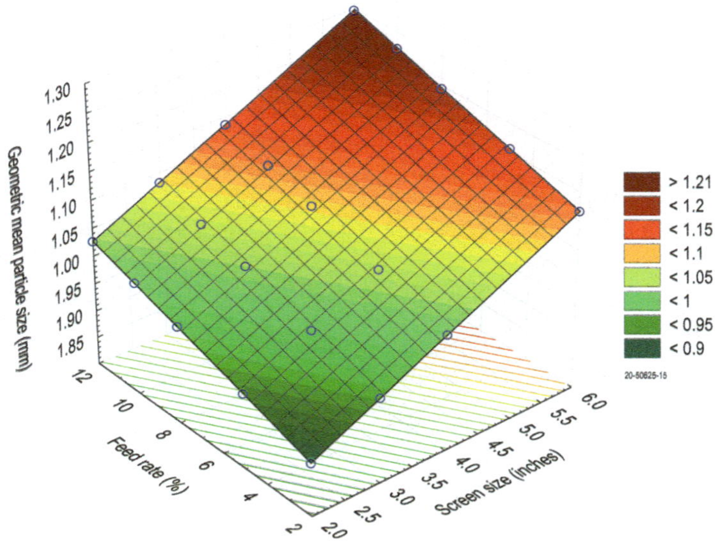

Fig. 4.17 Effect of feed rate and stage-1 grinder screen size on the geometric mean particle size at 20% bale moisture content

4.1.1.9 Grind Moisture Content

Surface plots were drawn to understand how the screen size in the stage-1 grinder, the feed rate, and the bale moisture content affect the final moisture content of the grind produced after fractional milling. Figure 4.18 indicates that bale moisture content has a significant impact on the final moisture content of the grind produced after fractional milling. At 25% bale moisture content and a 2% feed rate, the final moisture content of the grind was about 12.5%. This indicates that there is about 12–13% moisture loss during the fractional milling process. Figure 4.19 indicates that the feed rate did not have a significant impact on the final grind moisture compared to bale moisture content. At 25% bale moisture content and a 2% feed rate, the final moisture content was about >14% for a 3-inch screen. Figure 4.20 drawn for screen size and feed rate indicates that both the process variables had a significant effect on the grind moisture content. Similar plots were drawn for percent bypass using the equation in Table 4.1. The results indicate that increasing the screen size, moisture content, and feed rate have reduced the percent bypass of the material that has met the specifications. Further work is aimed at developing response surface models, which can help to understand the interactive effect of the process variables on the total grinding energy, geometric mean particle size, final moisture content of the grind, and percent bypass.

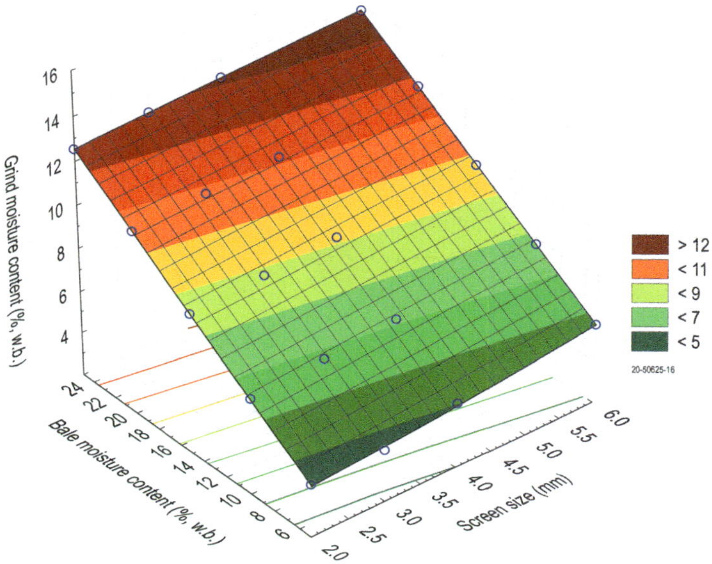

Fig. 4.18 Effect of stage-1 grinder screen size and bale moisture content on final grind moisture content at a 10% feed rate

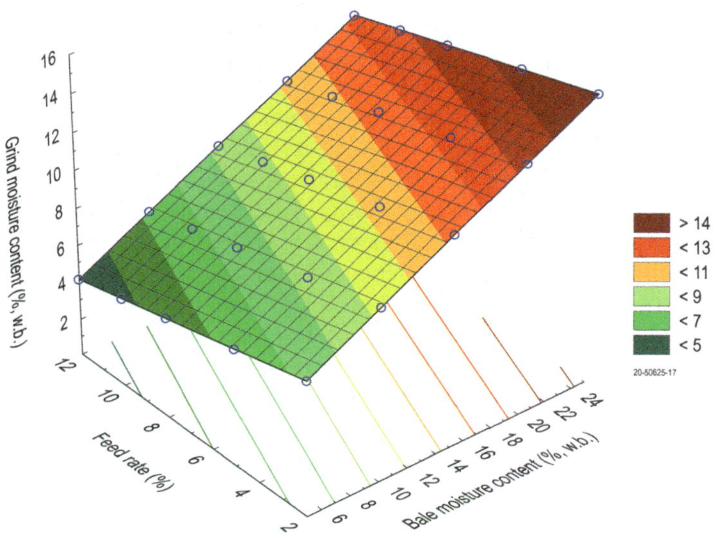

Fig. 4.19 Effect of bale moisture content and feed rate on the final grind moisture content for 3-inch screen size

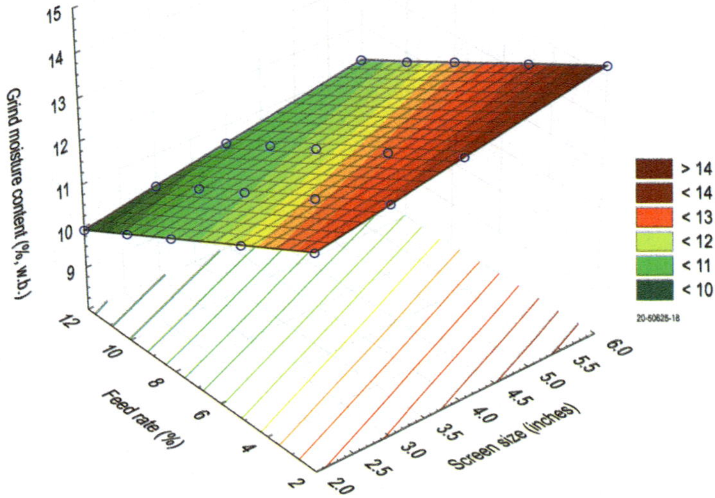

Fig. 4.20 Effect of feed rate and stage-1 grinder screen size on the final grind moisture content at a 20% bale moisture content

4.1.1.10 High-Moisture Pelleting and Low-Temperature Drying

One way to decrease drying costs and manage feedstock moisture content more efficiently is to use a high-moisture pelleting process coupled with a low-temperature drying method. High-moisture pelleting differs from current pelleting SOT by replacing the drying step (typically using a rotary dryer) after grinding with a low-temperature dryer (approximately 110 °C) for a short duration (typically several minutes) prior to pelleting. Instead, drying is an optional process that is inserted after pelleting. The main advantage of this method is that it provides an economical alternative to biomass drying. The combination of preheating, frictional heat generated in the pellet die, and cooling the pellets in a pellet cooler all help to partially dry the biomass, thereby removing about 10–12 points of moisture (e.g., from 30% down to about 18 to 20% w.b.) (Tumuluru 2014, 2015; Tumuluru et al. 2016). The produced pellets can be dried further, if necessary, to stable moisture content values of <10% (w.b.) using low-temperature and low-capital cost dryers, such as grain dryers (Kenney et al. 2013). In their studies, Lamers et al. (2015) indicated that a 5 ton/hr. rotary drier costs about $340,000, where a grain dryer costs about $33,000. Further, by moving the drying step to the end of the process, drying becomes optional and only used when high-durability, stability, and density are required, such as with long-distance transportation. INL believes that pelleting (even without drying) will provide a viable solution for feedstock handling and feeding problems, mainly when dealing with high- or variable-moisture feedstocks.

4.1.1.11 Fractional Milling and High-Moisture Pelleting Demonstration

The fractional milling, high-moisture pelleting, and low-temperature drying technologies not only help to reduce the cost of preprocessing by 50% as compared to current SOT for corn stover, but also turn the corn stover into an aerobically stable densified product (Kenney et al. 2013). In 2016, fractional milling and high-moisture pelleting were demonstrated for corn stover at the pilot-scale (Tumuluru et al. 2017). In September 2017, the integrated demonstration of fractional milling and higher moisture pelleting, as well as the further cooling of the pellets in a pellet cooler, was demonstrated using a 5 ton/hr. process demonstration unit (PDU) in the Biofuels Department at INL. In this demonstration, corn stover bales at 25% moisture content were processed based on the criteria provided previously in Fig. 4.7. The results indicated this process has a very significant effect on the cost of pellet production, as well as the quality of the pellets. This process has also helped to manage moisture content more efficiently. Combined fractional milling and high-moisture pelleting have reduced the moisture content in a corn stover bale from 25% down to about 12–13% (w.b.). These pellets were then cooled further in a pellet cooler to bring the moisture content down to <10% (w.b). The wet or high-moisture pellet durability was >97.5% in all the different process conditions tested. Another important variable that has improved the high-moisture pelleting process is the screen size of the stage-2 grinder. Commercial pellet producers recommend using a 1/4-inch grind in a hammermill to produce good quality pellets. But recent studies have indicated that using a 1/4-inch screen in the stage-2 grinder, especially at high-moisture content, increases the grinding energy significantly (Tumuluru et al. 2019). In its fractional milling and high-moisture pelleting demonstration project, INL changed the screen size of the stage-2 grinder to 7/16-inch and 1/2-inch to make grinding energy less sensitive to moisture content. This larger grind size was further sent to the pellet mill to produce pellets (Tumuluru et al. 2019).

Data regarding pellet quality produced after cooling is provided in Fig. 4.21, which indicates that the pellet has a smooth surface texture as compared to the pellets produced using conventional methods. In addition, the quality of the produced pellets was similar to the pellets made using a conventional method (Yancey et al. 2013). From the figure, it is evident that the pellets produced using high-moisture corn stover bales meet the quality specifications as defined by the Pellet Fuels Institute (PFI) and the European Committee for Standardization (CEN). Bulk density and durability are >630 kg/m^3 and > 97.5%, respectively (Tumuluru et al. 2019). The results indicate that the change in the particle grind size used for pelleting did not impact the quality of the pellets but resulted in a slight increase in pelleting energy (e.g., 40 kWh/dry ton for 1/4-in., 51 kWh/dry ton for 7/16-in., 62 kWh/dry ton for 1/2-in.). A full-blown techno-economic analysis (TEA) was conducted based on this data in terms of throughput, energy consumption, and capital costs of fractional milling, high-moisture pelleting, and cooling. The results indicate there is about a 65% reduction in pellet production costs as compared to the conventional pelleting method followed by industry, as observed in Fig. 4.22.

Fig. 4.21 Corn stover pellets produced using novel preprocessing technologies developed by INL (corn stover bales were processed using commercial-scale grinders and pellet mills available with Biomass National User Facility, INL)

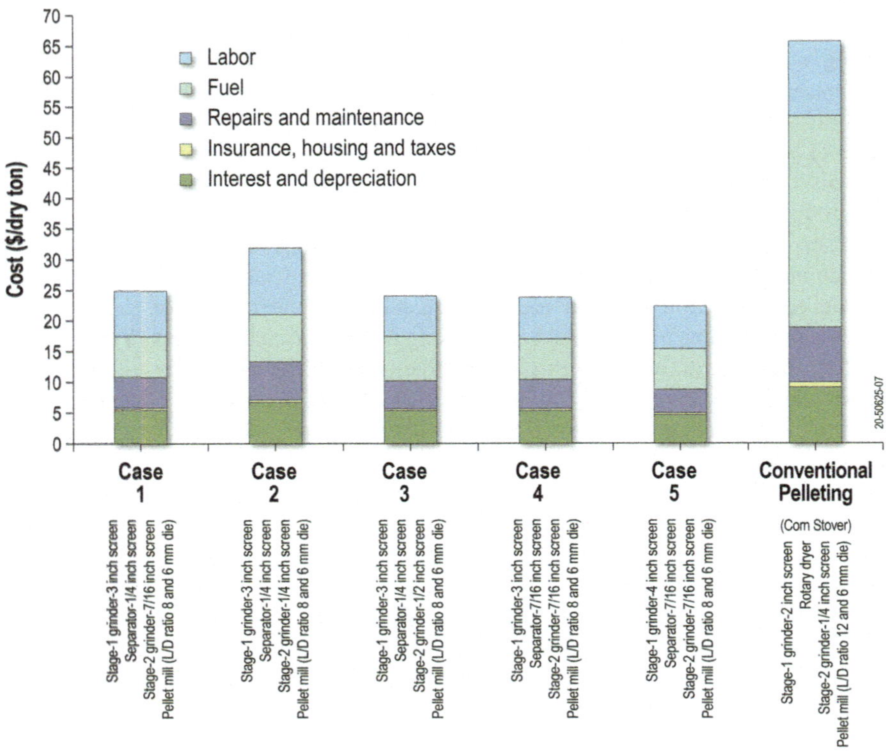

Fig. 4.22 Techno-economic analysis of fractional milling and low-temperature drying using the data produced using BFNUF (Biomass Feedstock National User Facility)

INL also observed that the length to diameter (L/D) ratio of the pellet die and the moisture content of the corn stover grind during high-moisture pelleting has a significant impact on pellet quality, especially bulk density and durability. Figures 4.23, 4.24, and 4.25 show the effect of L/D ratio and corn stover grind moisture content on bulk density, durability, and pelleting energy consumption, respectively. This data was produced using a pilot-scale ring die pellet mill (1 ton/hr.). The results indicate that bulk density, durability, and pelleting energy are influenced by both the moisture content of the biomass and the L/D ratio of the die. A higher L/D ratio and a lower corn stover moisture content increase the density and durability, whereas lower density in the range of 25–30 lbs./ft^3 is achievable at a higher moisture content and a lower L/D ratio. The high-moisture pelleting process is so flexible that the process can be optimized to produce pellets that are more desirable for transportation and conversion process requirements.

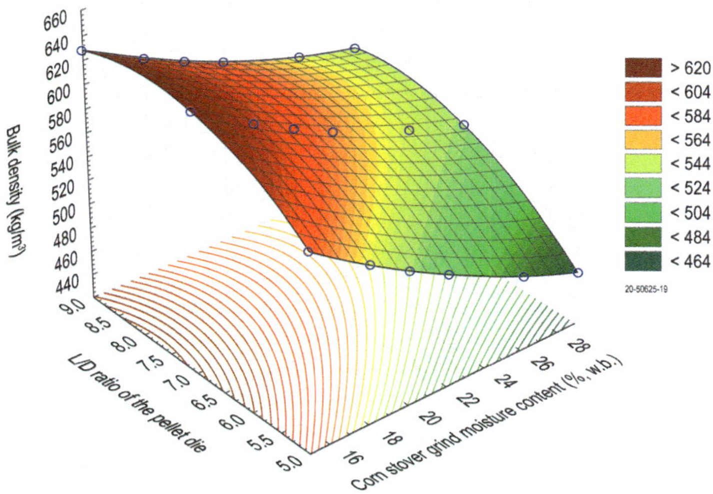

Fig. 4.23 Pellet die L/D ratio and grind moisture effect on bulk density

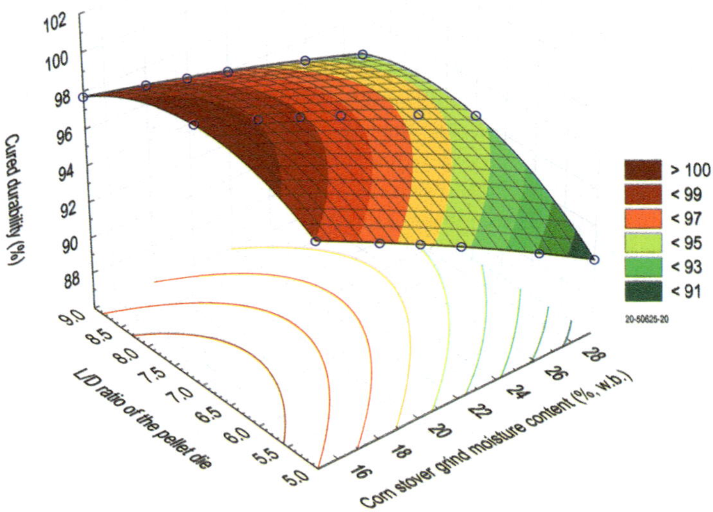

Fig. 4.24 Pellet die L/D ratio and grind moisture content effect on cured durability

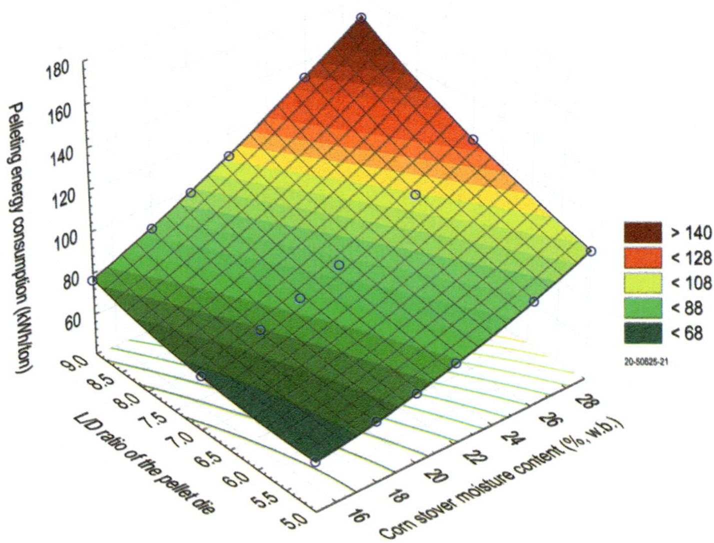

Fig. 4.25 Pellet die L/D ratio grind moisture content effect on pelleting energy

4.1.1.12 TEA of Fractional Milling, High-Moisture Pelleting, and Low-Temperature Drying

Figure 4.26 shows how grinding energy changes by using the stage-1 and stage-2 grinders. It also shows how moisture content impacts grinding energy. The conventional preprocessing cost to produce a densified product, such as a pellet, is given in Table 4.3. This table indicates that drying and stage-1 grinding account for the majority of the costs as compared to stage-2 grinding and pelleting. Figures 4.8 and 4.9 previously described the grinding costs for stage-1 and stage-2 grinders using the conventional method followed by industry. Table 4.4 indicates the grinding costs with stage-1 and stage-2 grinders with a separator (Tumuluru and Yancey 2018). It is clear from this table that changing the screen size from 25.4 mm to 6.35 mm in the stage-2 grinder increases grinding costs by almost six times. In addition, moisture content has a significant effect on grinding costs as well.

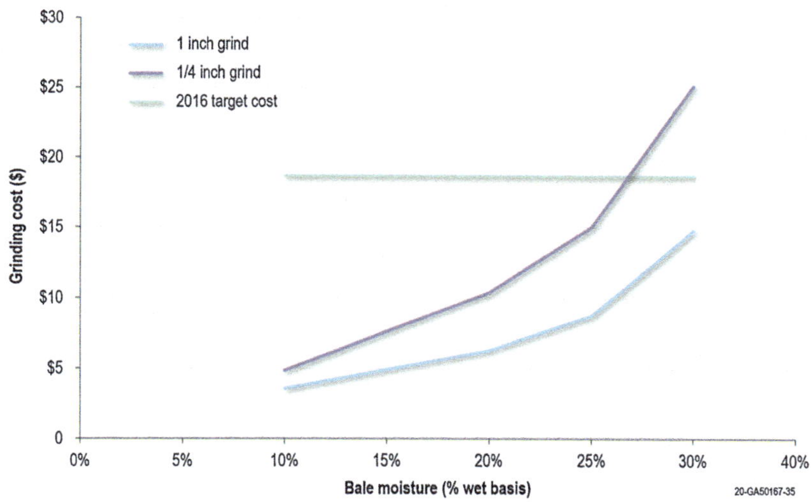

Fig. 4.26 Corn stover grinding cost using a stage-1 and stage-2 grinder for different moisture content values and screen sizes (derived from Tumuluru and Yancey 2018)

Table 4.3 Conventional preprocessing systems cost (derived from Kenney et al. 2013)

2013 State of technology (2011 $/dry T)	
Grinder 1	16.80
Grinder 2	11.60
Drying	15.20
Densification	7.70
Totals	51.30

Table 4.4 Grinding cost demonstrated in 2016 using stage-1 and stage-2 grinders and a separator (derived from Tumuluru and Yancey 2018)

Cost (in 2014 $/dry ton)				
Grinding conditions (bale moisture and screen size)	10% M.C. 76.2 mm screen in stage-1 and 25.4 mm screen in stage-2	30% M.C. 76.2 mm screen in stage-1 and 25.4 mm screen in stage-2	10% M.C. 76.2 mm screen in stage-1 and 6.35 mm screen in stage-2	30% M.C. 76.2 mm screen in stage-1 and 6.35 mm screen in stage-2
Stage 1 grinder costs	$2.55	$11.67	$2.55	$11.67
Separator	$0.37	$0.37	$0.37	$0.37
Stage 2 grinder costs	$1.06	$3.09	$2.33	$13.42
Total	$3.98	$15.13	$5.25	$25.45

4.1.1.13 High-Moisture Pelleting

The TEA shown in Table 4.5 provides the cost of pelleting production using the high-moisture pelleting and low-temperature drying process. The cost to produce corn stover pellets using a rotary dryer and low-moisture pelleting is $22.50/dry ton, as shown in Table 4.5. The high-moisture pelleting conducted on corn stover at 20% (w.b.) cost about $9.38/dry ton, which is also shown in Table 4.5. The high-moisture pelleting reduces pelleting costs by about $12.92/dry ton—a reduction of 40% as compared to the conventional pelleting process. In the conventional case, the dryer is a rotary dryer, and the high-moisture pelleting case uses a crossflow grain dryer, both fueled by natural gas at $7.55/MMBTU (million BTU). Additionally, the type of dryer has changed from a traditional grain dryer with a capital cost of $270,000, which is more efficient, has a less capital-intensive crossflow grain dryer, and a capital cost of $35,009 (Lamers et al. 2015). The change from a rotary dryer (conventional case) to a crossflow grain dryer has significantly reduced costs (Lamers et al. 2015). Table 4.6 indicates the cost to preprocess a bale with a 30% w.b. moisture content using fractional milling with a 3-inch screen in the stage-1 grinder and a 1/4-inch screen in the stage-2 grinder, followed by high-moisture pelleting and low-temperature drying. The table clearly indicates that the cost to produce pellets using corn stover bales at a 30% w.b. moisture content using fractional milling, high-moisture content, and low-temperature drying has reduced by about 50% as compared to the traditional method (Lamers et al. 2015). Table 4.7 indicates the cost of pellet production when the screen size of the stage-2 grinder is switched from 1/4-inch to 7/16-inch. It is also clear from the table that increasing the stage-2 grinder screen size to a 7/16-inch screen reduces the pellet production cost by about $4/dry ton.

Table 4.5 TEA of conventional and high-moisture pelleting process (derived from Tumuluru and Yancey 2018)

Case studies	Densifier ($/dry ton)	Dryer ($/dry ton)	Total ($/dry ton)
Conventional (Kenney et al. 2013)	7.70	15.20 (from initial moisture of 30% w.b. to final moisture of 12% w.b. using a rotary dryer)	22.50
HMPP with a grain dryer	8.31	1.27	9.58

Table 4.6 Overall costs to produce pellets using fractional milling, high-moisture pelleting, and low-temperature drying (derived from Tumuluru and Yancey 2018)

Corn stover feedstock (30% w.b.)	Cost ($/ton)	Energy (kWh/ton)	Moisture loss (% w.b.)	Mill throughput (dry ton/hr.)
Stage-1 (3-in. grind)	10.78	15.9	30	3.0
Stage-2 (1/4-in. grind)	8.52	35.3 (54.3 w/ 35% bypass)	25	2.2
Separator screen size (1/4-in.)	0.21		1	5
Pelleting	8.31	73	19	4.8
Drying (grain dryer)	1.27	50	$\leq$10	5
Total	29.09			

Table 4.7 The effect of stage-2 grinder screen size on pellet production costs (derived from Tumuluru et al. 2017)

Stage-2 grinder screen size	Pellet production cost ($/dry ton)
Case-1 (stage-2 grinder fitted with 1/4-in. screen size)	29.09
Case-1 (stage-2 grinder fitted with 5/16-in. screen size)	27.66
Case-1 (stage-2 grinder fitted with 3/8-in. screen size)	27.17
Case-1 (stage-2 grinder fitted with 7/16-in. screen size)	25.35

Note: Preprocessing cost targets achieved with fractional milling, high-moisture pelleting, and low-temperature drying using pilot-scale systems

4.1.1.14 Comparison of Conventional Pelleting, High-Moisture Pelleting, and Conventional Briquetting Processes

Figures 4.27 and 4.28 indicate the total energy consumption and overall costs, respectively, for making pellets using a conventional pelleting process (CPP), a high-moisture pelleting process (HMPP), and a conventional briquetting process (CBP) where the raw material is dried to 10% (w.b.) before briquetting with a bigger hammer screen size of 1/2-inch (Tumuluru et al. 2014). The TEA indicated that even though the energy consumption of the conventional briquetting process is less as compared to the CPP, the total costs are higher by about 9.8% (w.b.) for briquettes as compared to pellets. This increase in the total cost can be attributed to higher capital costs involved in a briquetting press due to limitations in their production capacities,

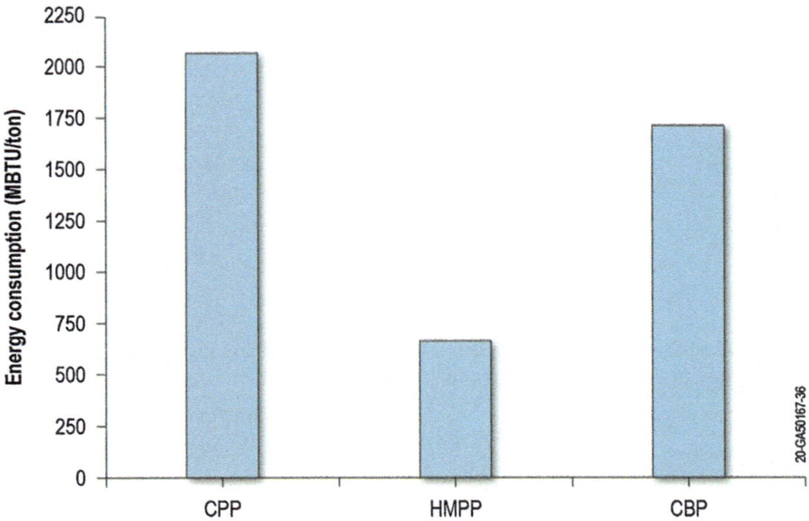

Fig. 4.27 Comparison of energy consumption for the different densification processes (derived from Tumuluru et al. 2014)

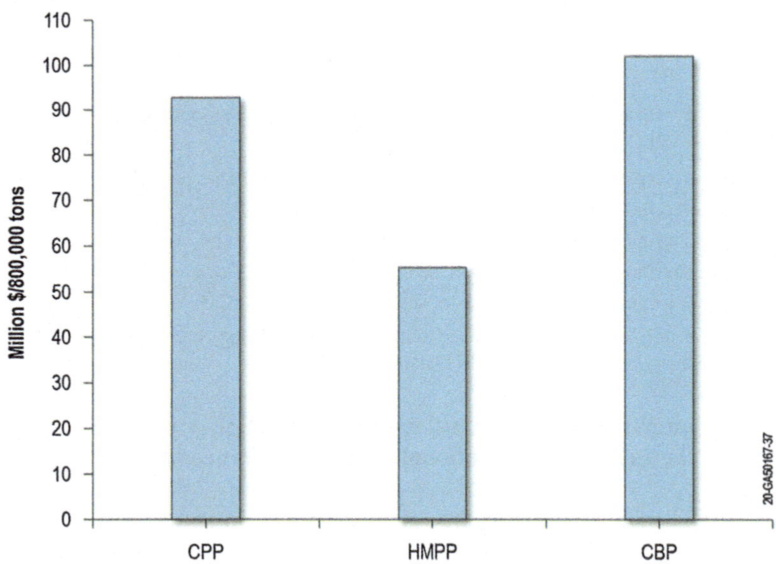

Fig. 4.28 Comparison of costs for the different densification processes (derived from Tumuluru et al. 2014)

which equates to 2 ton/hr. as compared to 5 ton/hr. for a pellet mill. The TEA also indicated that a significant reduction in energy consumption and costs could be achieved using the high-moisture pelleting process as compared to the CPP and CBP.

4.1.1.15 Novel Technologies to Reduce the Biomass Preprocessing Costs Using Low-Temperature Drying and Dewatering Technologies

According to Roos (2013), biomass drying helps to improve combustion efficiency, which results in higher steam production, a reduction in emissions, and an increase in boiler efficiency. For example, with wood chips at 45% (w.b.) moisture content, the maximum boiler efficiency achieved using a standard system is about 74% (w.b.), whereas if dried woodchips at about 10% (w.b.) are used in the same system, the system efficiencies go up to about 80% (w.b.). This increase in system efficiencies can also result in higher steam production (Roos 2013). The other challenges in burning high-moisture wood chips are to keep the flame burning, whereas dry fuels tend to burn hotter and more evenly for longer, as well as to reduce ash quantity. Another significant advantage in drying biomass fuels is that it reduces fuel transportation costs and keeps the fuel aerobically stable (i.e., minimizing dry matter losses). The moisture content of woody and some herbaceous biomasses after harvesting is higher than 30% (w.b.). Table 4.8 provides a list of the high-moisture content values for some of the more common woody, herbaceous, food waste, and manure biomasses (Roos 2013), such as apple pomace, cherry, pine chips, pulp, and paper mill sludges, corn cob, cow manure, and MSW.

Table 4.8 Moisture content by weight of several biomass feedstocks as received

Feedstock	Moisture content (%, w.b.)
Food wastes	
Fruits	
Apple pomace	72
Cherry	37.8
Orange peels	10.8
Melon shells	27.6
Nuts	
Black walnut Shell	11.6
Peanut hulls	9
Peanut skins	8
Vegetables	
Wet potato wastes	86
Forest products	
Pine chips	54.6
Pine sawmill waste	11.3
Construction waste	12–17
Pulp & Paper Mill Sludge	50–70
Agricultural waste	
Rice husks	10 (as received) 8 (air-dried)
Corn cob	10–43
Soy hulls	9

(continued)

Table 4.8 (continued)

Feedstock	Moisture content (%, w.b.)
Food wastes	
Lactating cow manure	88 (as excreted) 98 to 99.7 (from milk house or parlor)
Fresh and marine biomass	
Municipal waste	>95
Sewage sludge-biosolids	90–97
Septage-biosolids	98
Municipal solid waste	12–32
Other	
Fish waste	76
Fruits and vegetables waste-grocery store	88

4.1.1.16 Low- and Medium-Temperature Drying Technologies

Low- and medium-temperature drying technologies are gaining a lot of attention as compared to high-temperature biomass drying. This is primarily due to: (a) lower capital costs; (b) low-operational costs; (c) low-volatile organic emissions during drying; (d) fewer particulate emissions; and (e) fewer fire hazards. As indicated in Table 4.7, most of the biomass used for biofuels production at the time of harvest results in high moistures. However, these also need to be dried to even less moisture content (typically up to 10%, w.b.) in order to effectively mill and further densify. High-moisture content values in the biomass have a negative impact on grinding as most grinding systems currently used by industry are sensitive to moisture. There are two methods of drying—passive and active. The major drawback of passive drying is that it takes more time and limits the availability of biomass for biorefineries. In addition, high-temperature drying—such as using a rotary dryer—commonly used by biorefineries, paper, and pulp factories, and lumber mills are very energy-intensive, requiring a high-quality heat. Studies conducted by many researchers have suggested that drying is the major energy consumer in the pellet production process (Tumuluru 2015, 2016). All these issues resulted in biorefineries to look for alternative ways to dry biomass at lower costs.

4.1.1.17 Low-Temperature Drying of Biomass

Compared to high-temperature drying processes (i.e., rotary dryers), low-temperature drying methods are less expensive and require less equipment and less external energy input. The commonly used low-temperature dryers are belt dryers, where fans blow the drying medium in either an upward or downward direction (Tumuluru and Yancey 2018) as these dryers operate at temperatures <100 °C, low-grade fuel—such as hog fuel—can be used to generate heat. A typical conveyor or belt dryer used by industry is shown in Fig. 4.29. The typical

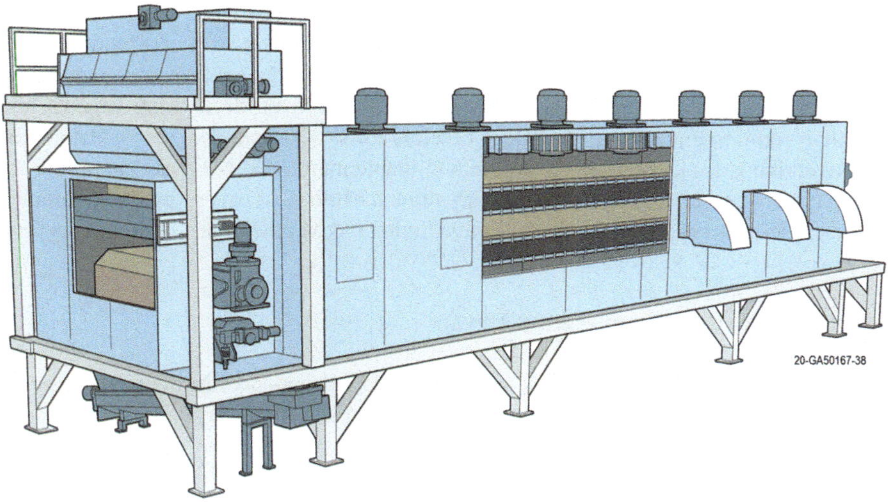

Fig. 4.29 Typical conveyor or belt dryer for biomass drying

temperatures used for the drying of biomass using low-temperature methods are somewhere between 60 and 80 °C. Examples of low-temperature dryers are cabinet dryers, grain dryers, and belt dryers (Lamers et al. 2015).

The major advantages of low-temperature drying are: (a) decreasing air pollutants from the drying process; and (b) energy-efficiency due to low-temperature dryer settings. The increase in efficiency is due to less heat loss through the dryer due to a smaller temperature gradient. The other advantage of low-temperature drying is low fuel consumption. This gives the flexibility to utilize low-potential heat sources—like sun, heat pumps, and waste heat—from production. Low-temperature is widely used in food processing industries—especially in the drying of fruits, vegetables, and grains. As the components of food and other physical, color, and textural properties are sensitive to higher temperatures, low-temperature drying is the most preferred option (Nagaya et al. 2006; Zlatanović et al. 2013). In Europe, low-temperature drying technologies are gaining more importance in biomass drying due to environmental and cost concerns. In their study on methods to produce durable pellets at lower energy consumption, Tumuluru et al. (2016) indicated that low-temperature drying methods have several advantages: (a) higher efficiency; (b) a reduced fire hazard; (c) no need of high-quality heat; (d) reduced VOC emissions; (e) reduced particulate emissions; (f) no high-clay agglomeration or sticky biomass; and (g) easy access for maintenance (Tumuluru et al. 2016). Currently, low-temperature dryers such as grain or belt dryers are available at a commercial scale (e.g., >5 ton/hr.). These dryers are so flexible that they can be used for a variety of products. Examples of companies currently selling these commercial-scale belt dryers include Haarslev Industries A/S, Bogensevej DK, and Perry of Oakley, Devon, UK. Conveyor dryers are available in single- and multi-layer formats. The footprint of single-pass conveyor dryers is typically larger than a comparably sized rotary dryer. Multi-pass conveyors are arranged one above the other with material cascading down from the upper conveyors to the lower conveyers, saving a considerable amount of space. Multi-pass dryers are very common in many industries due to their small footprint and lower costs. The capital cost of conveyor and rotary dryers is often comparable. However, a conveyor dryer may require less ancillary equipment for the treatment of emissions. As such, for new installations, the overall costs may be less. Operation and maintenance (O&M) costs are higher for these types of dryers as compared to rotary dryers. Multi-pass dryers are more complex than single-pass dryers. As such, they have higher O&M costs than single-pass dryers.

4.1.1.18 Medium-Temperature Dryers

Cascade dryers, flash dryers, and superheated steam dryers operate at medium drying temperatures. These dryers are widely used for biomass drying. The biomass material in these dryers is introduced into a hot air stream current. The air is initially carried upward and further cascades back to the bottom where it is lifted again. Flash dryers are generally cost-effective only at larger scales. Electricity use by flash dryers is greater than that of other dryer types because high airflows are required to keep the material suspended. Flash dryers require a small particle size, so shredders may be required, which leads to increased electrical use. Flash dryers have a small footprint. On the downside, they are subject to corrosion and erosion problems. They also have a fire risk following the shutdown of the dryer.

Cascade dryers typically operate at intermediate temperatures of lower and higher temperature dryers. In the case of flash dryers, the feedstock is suspended in an upward flow of the drying medium, which is generally flue gas. These dryers are used for a wide variety of materials. The particle needs to be smaller in size to suspend them in the flue gas in a flash dryer. Superheated dryers are like flash dryers, but instead of flue gas, steam is used in the medium. In steam dryers, the temperature of the steam is normally kept above the saturation temperature in order to avoid steam condensation, thereby transferring the only sensible heat.

These dryers can be operated in a closed-loop with low-pressure steam from the dryers where the steam can be reheated and injected back into the dryer. As there is a large quantity of steam that leaves the dryer, the excess steam can be recovered and used in other processes where a lot of this energy can be recovered. There are no emissions associated with these dyers, as well as no fire hazards. Therefore, the footprint of these dryers is smaller as compared to high-temperature rotary dryers.

4.1.1.19 Dewatering Technologies to Manage Moisture More Efficiently

Among the various unit operations in the pellet production process, drying is the most energy-intensive. As investigated by many researchers, drying takes about 65–70% of the total energy. The high-energy consumption of drying is due to the high-quality heat requirement. One way to reduce drying costs is to design the pelleting process with a low-temperature energy supply. There is a lot of waste energy in the biorefinery that can be used for drying, which can have a significant impact on pelleting costs. According to Frodeson et al. (2019), if the waste energy from saw and pulp mills can be used as an energy source for biomass drying, pellet production costs can be reduced significantly.

Frodeson et al. (2019) state that biomass drying is divided into three phases. During the first and second phases, the diffusion of water occurs between the air moisture interface, which is influenced by the thermal energy of the dryer. In the last phase, the moisture is transported through the biomass, which is based on concentration gradients. The last phase takes a lot of energy to drive the moisture. Industrial pelleting data has indicated that two-step drying is more efficient in the pellet production process. According to Ståhl and Berghel (2008), recirculating the drying gases makes the drying process more energy-efficient due to the recovery of more energy in the condenser. The other way to reduce drying energy is to mechanically dewater the biomass before thermal drying. Most of the surface moisture or loosely bound moisture can be removed using the mechanical dewatering process. Mechanical dewatering can increase the overall efficiency of the moisture removal process. Dewatering the biomass is financially gainable, saves energy, and provides a sustainable solution for the environment.

Drying beds, filters, screens, presses, and centrifuges are just a few of the dewatering pieces of equipment used by industry. The type of dewatering used for specific applications depends on feedstock type and the initial moisture content of the feedstock. According to Roos (2013), mechanical dewatering systems can reduce moisture content significantly. Some of the more passive drying methods, such as filter bags, are impervious to rain, but allow moisture to seep out, which can help to reach moisture content values of 30% (w.b.). However, the major challenge is it takes longer times to achieve–approximately two to three months. The different types of mechanical presses typically used for dewatering applications include the belt filter, a V-type, the ring, the screw, and the drum presses (Roos 2013). In a belt press, the material is pressed against a porous belt and is further passed through the rollers to squeeze out the moisture. Belt presses are commonly used in many industries, especially in wastewater treatment plants. They contain three solid and liquid separations as follows: (a) the gravity zone, where water in the feed is allowed to drain freely through the porous belts; (b) the wedge zone, where the solids are prepared for pressure applications; and (c) the pressure zone, where solids are subjected to medium and high pressure. However, the major challenge in using this process is the frequent need for replacement belts due to wear and tear (e.g., typically, belts only have a lifespan of 2 to 3000 hrs.). Figure 4.29 shows a typical belt dryer used for the dewatering application (Perinpanayagam 2013).

A centrifuge system works by moving the material into a stationary tube that is further fed into a rotary bowl, which contains an accelerating inlet rotary, as shown in Figs. 4.30 and 4.31. The rotary turns at a speed of 3400 revolutions per minute (rpm), creating high-centrifugal forces pushing the solids to the bowl wall. The conveyor in the centrifuge rotates in the opposite direction of the rotor, which further moves the solids to the wall (Perinpanayagam 2013).

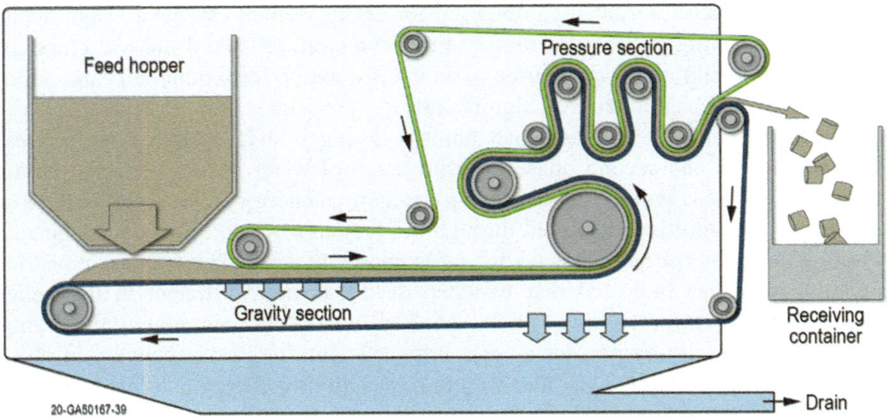

Fig. 4.30 Belt filter press schematic

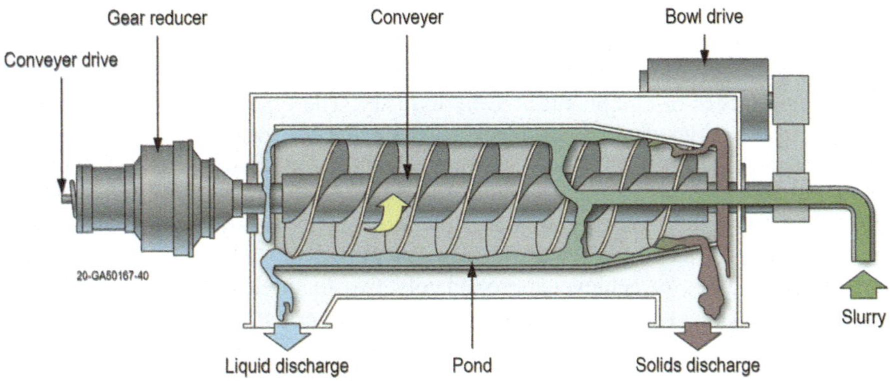

Fig. 4.31 Typical centrifuge thickening and dewatering system (derived from EPA 2000)

A screw press consists of a hollow cylindrical screen mounted at an angle or horizontally, which is rotated at low speeds, as is shown in Fig. 4.32. As the material is conveyed through the barrel length, the tapered screw helps to squeeze the water from the solids. The low-concentration solids (i.e., 1–3%) are dewatered (i.e., 15–30%) by gradual application of pressure and polymer addition. The outputs of the screw press are filtrates and dewatered solids.

The plate and frame filter press (PFFP) is one of the oldest solid separation methods producing the highest concentration of solids, as observed in Fig. 4.33. It consists of chambers created by a series of recessed plates arranged vertically side by side. These plates are supported on a frame connected to the stationary feed head. The chambers are filled with the material through the center hole. With an increase in pressure, the free water drains through the filter cloth, and solids are retained in the chamber. Typically, the feed pumps should develop high pressures of 100–225 pounds per square inch (psi). Once the pump reaches the desired pressure, the pumps are switched off.

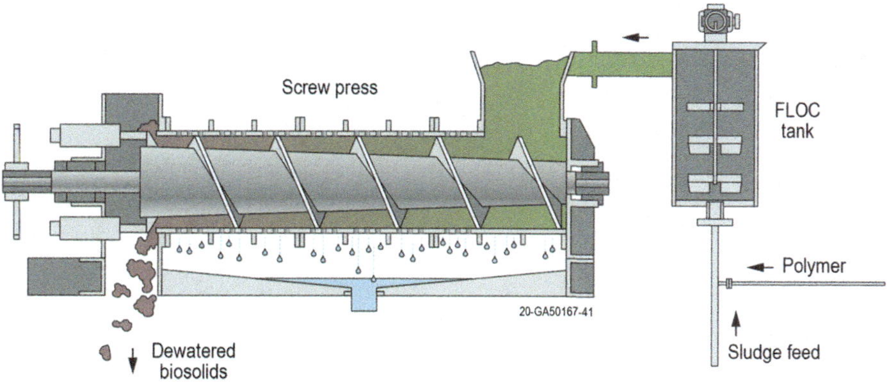

Fig. 4.32 FKC screw press schematic diagram

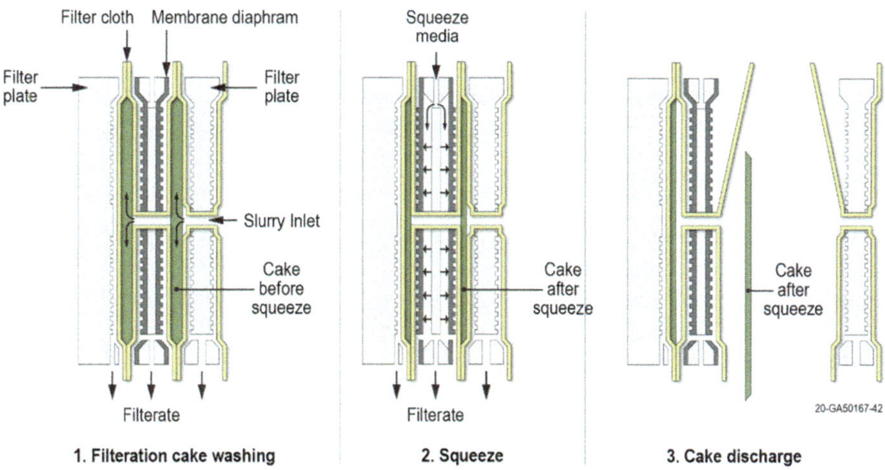

Fig. 4.33 Schematic diagram of the PFFP operation

The drum press has a perforated drum with a revolving press inside of it that presses it against the perforated drum. This press is commonly used for biomass materials, such as hog fuel and bark (Roos 2013). The rotary press uses backpressure to produce sludge cake, as shown in Fig. 4.34. The sludge is fed through the inlet into the space between the two parallel plated screens and the rotor. The backpressure applied at the discharge end pushes the material to dewater. The filtrate passes through the screen as the material continues to dewater as it moves through the channel. Due to the slow moving of the screen, the frictional force is developed that also helps to dewater. Most of these presses are currently used in biosolids processing. These technologies need to be modified for use in biomass applications. Drinor Corporation in Sweden developed a mechanical continuous dewatering press (CDP) specifically for biomass. This press can be used as a pretreatment that can

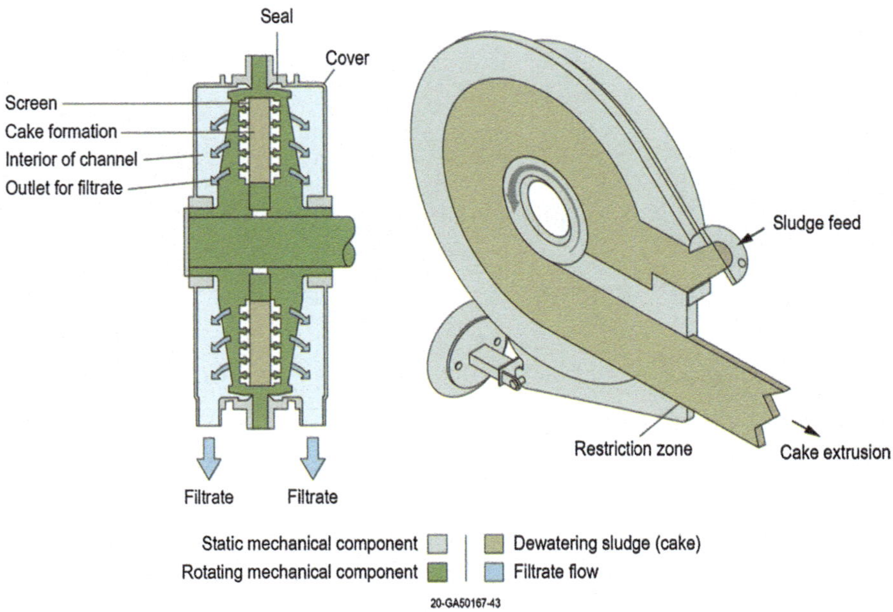

Fig. 4.34 Rotary press schematic diagram

increase the efficiency of the drying process and reduce energy consumption. The CDP prevents the reabsorption of water and delivers a product with a constant moisture content of about 30% (w.b.). Biomass, such as woodchips and sawdust, can be successfully used for efficient surface water removal.

In general, moisture removal using thermal methods is more energy-intensive as compared to the mechanical dewatering process. According to Liu and Haygreen (2007), the mechanical dewatering process helps to remove the free moisture under hydrostatic pressure and is more energy-efficient than traditional thermal drying. Studies conducted by Laurila et al. (2014) on a laboratory-scale compression dryer indicated that compressive drying is an effective solution for removing the moisture up to 30% (w.b), which is about 7–12% above the fiber saturation process. In their studies on roller compression with thermal drying, Yoshida et al. (2010) discovered that using this method saves about 58% of the energy than by just employing the thermal drying method alone. Studies conducted by Holmberg and Stenström (2014) indicated that dewatering, in combination with thermal heating, makes water removal more efficient. Frodeson et al. (2019) did a case study to understand how mechanical dewatering combined with the thermal method helps to reduce the energy consumption of the pellet production process, as shown in Fig. 4.35. Table 4.9 indicates the energy balance for pellet production by using the continuous drying system alone, as well as combining it with dewatering (Frodeson et al. 2019). It is clear from the table that electric power is much higher when dewatering is used. In addition, Table 4.9 indicates that the throughput of the system is higher when the

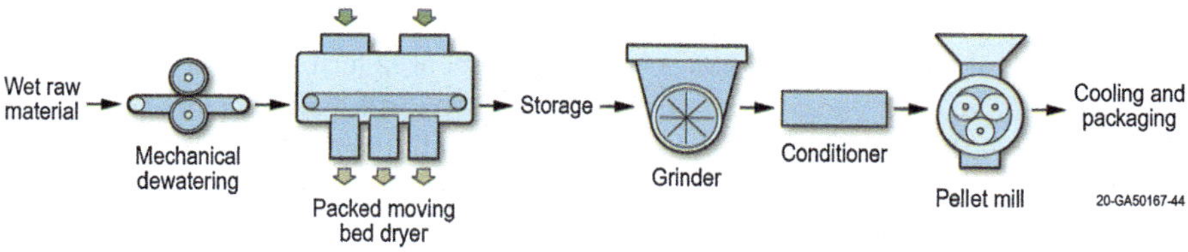

Fig. 4.35 Mechanical dewatering press and packed moving bed (PMB) dryer in a pellet production chain (Frodeson et al. 2019)

Table 4.9 Inlet data for continuous drying system and dewatering + continuous drying system (Frodeson et al. 2019)

	Packed moving bed (PMB) dryer	Continuous dewatering press + packed moving bed (CDP + PMB)
Average inlet air temperature	75 °C	75 °C
Heat power provided to the heat exchangers	10.1 MW	5.1 MW
Electric power supplied to the fans in the dryer	0.450 MW	0.22 Mw
Thermal losses from packed moving bed (PMB) dryer	10% of inlet power	10% of inlet power
Inlet mass flow of sawdust	18.53 ton/hr.	18.58 ton/hr.
Inlet sawdust moisture content	52.0%	52.0%
Outlet sawdust moisture content	12.5%	12.5%
Outlet sawdust moisture from continuous dewatering press (CDP)	–	36.7%
Current load CDP	–	209A
Voltage CDP	–	400 V

dewatering and drying system is employed. Novel and energy-efficient water management technologies are needed in order to reduce the drying costs, especially for higher moisture biomass.

4.1.1.20 Advanced Grinding Concepts

The biomass after harvesting is in the form of a baled, log, or chip form to store and transport. Hammermills are typically used for size reduction of low-moisture biomass. When the moisture changes, the grinding energy and physical properties of the biomass are impacted, such as particle size and distribution, mean particle size, and bulk density. Biomass grinding is completed in two stages. In the first, the typically 30% (w.b.) biomass is ground in a hammermill fitted with larger-size screens. In the second, the ground biomass is further ground with smaller-size screens that are less than 1-inch. Figures 4.8 and 4.9 previously showed the energy consumption of the hammermill with changes in moisture content. From the figures, it is very clear that the hammermill works well for lower-moisture content in terms of energy consumption, throughput, and particle size distribution (Tumuluru and Yancey 2018). Still, most of the biofuels processing industries are currently using hammermill technology for grinding high- and low-moisture woody and herbaceous biomass. According to Tumuluru and Yancey (2018), hammermills are sensitive to moisture, and they work for low-moisture woody and herbaceous biomass. This necessitates the need to develop new grinding technologies that are less sensitive to moisture content. New grinding technologies based on shearing and cutting are being investigated to determine their suitability for biomass grinding. Forest Concepts, LLC, developed Crumbler model M24, which is a biomass size reduction technology based on shearing (Tumuluru and Yancey 2018), which employs 70-cm-wide interlocking disks. This technology is less sensitive to moisture content. The spacing between cutters has a great impact on the resulting particle size distribution, where tighter particle size distribution can be achieved.

The use of a vortex mill, which is widely used in ore processing, is also currently being tested for use on biomass. Figure 4.36 indicates the energy consumption of the different grinding systems at different moisture content values. The vortex or collision mill required more energy to process biomass, especially high-moisture biomass, as compared to the other two technologies. The shearing technology performed the best on high-moisture corn stover. The hammermill performed better than the other two technologies for woody material. Integrating the new grinding technologies with the pelleting process can help to reduce the cost of pellet production significantly. Studies conducted by Bitra et al. (2009) on understanding the grinding characteristics of switchgrass, wheat straw, and corn stover using a knife mill indicated that the total specific energy decreased gradually by 55%, 49%, and 75%, respectively, with an increase in mass feed rate from 2–11 kilograms per

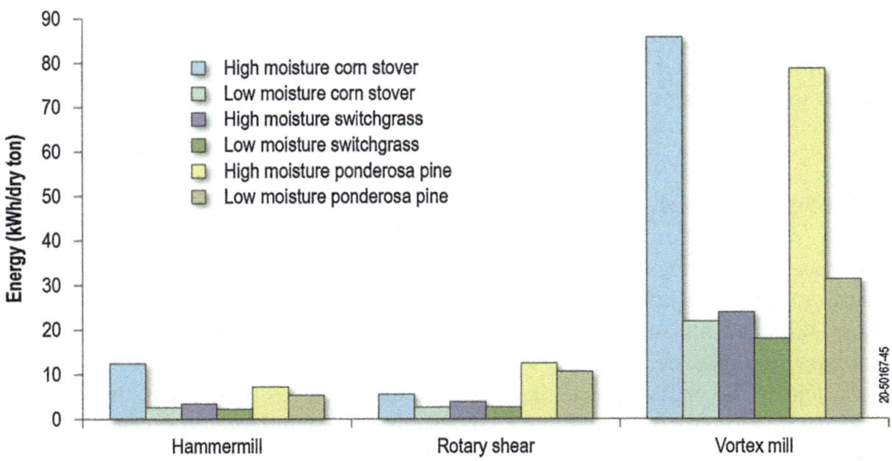

Fig. 4.36 Specific grinding energy for three grinder types

minute (kg/min), 2–9 kg/min, and 2–7 kg/min for switchgrass, wheat straw, and corn stover, respectively. For the same operating conditions, the effective specific energy increased marginally by 11% and 4% for switchgrass and wheat straw, respectively, but decreased marginally by 7% for corn stover. Lower specific energy at higher feed rates can be attributed to grinding of more material in a specified time and because of better utilization of the full length of the rotor. So, it is recommended to run these grinders at higher feed rates to lower their energy consumption. However, these higher feed rates can have an impact on physical properties, such as density and particle size distribution. In the case of using a knife mill with a screen size of 24.40 mm and a rotor speed of 250 rpm, the feed rates of 7.6, 5.8, and 4.5 kg/min, respectively, provided total specific energy rates of 7.57, 10.53, and 887 kWh/Mg, respectively. These authors concluded that screen size, mass feed rate, and rotor speed had a significant effect on the total specific energy consumption of the grinding process. Selecting the right grinding technology based on the type of feedstock can have a substantial impact on the pellet production cost. Further research is needed to manage moisture efficiently and to reduce the energy consumption of the various preprocessing unit operations to reduce the pellet production cost.

4.2 Conclusions

Currently, preprocessing contributes to about 35–40% of the overall feedstock costs supplied to the biorefineries. The major energy consumers are drying biomass to 10% (w.b.) to make it aerobically stable, as well as grinding the high-moisture biomass using hammermills. New grinding, pelleting, and drying technologies are gaining importance as these technologies help to reduce the total energy consumption of preprocessing, and in turn, significantly reduce costs. INL has developed a fractional milling, high-moisture pelleting, and low-temperature process that helps to minimize pellet production costs by more than 50% as compared to the methods currently followed by industry. Typically, hammermills are sensitive to moisture and screen size, whereas higher moisture and smaller screen size exponentially increase grinding energy. New grinding technologies, such as the rotary shear, are gaining confidence throughout the industry as they are less sensitive to moisture. In addition, studies conducted by researchers have indicated that herbaceous biomass can be pelleted at even bigger grind sizes (e.g., 11.11 mm) as compared to the screen size currently employed by industry, which is typically 6.35 mm. This reduces grinding energy consumption, especially when high-moisture biomass is used for pelleting. In biomass preprocessing, the major energy consumer is moisture removal. Thermal methods to reduce the biomass moisture, like using a rotary dryer that is now commonly used, remain one of the major bottlenecks in reducing overall preprocessing costs. Researchers are also integrating dewatering and low-temperature drying technologies in the pellet production process to reduce the overall total energy consumption for pelleting biomass.

References

Bell TA (2005) Challenges in the scale-up of particulate processes-an industrial perspective. Powder Technol 150:60–71

Bengtsson P (2004) The release of hydrocarbons from softwood drying: measurement and modeling. Maderas: Ciencia y Tecnología 6(2):109–122

Bitra VSP, Womac AR, Igathinathane C, Miu PI, Yang YT, Smith DR, Chevanan N, Sokhansanj S (2009) Direct measures of mechanical energy for knife mill size reduction of switchgrass, wheat straw, and corn stover. Bioresour Technol 100:6578–6585

Frodeson S, Berghel J, Stahl M, Granstrom K, Romlin C, Thelander A (2019) The potential for a pellet plant to become a biorefinery. PRO 7(4):233

Holmberg A, Stenström S (2014) Dewatering and drying of bark. Int J Energy Eng 4:8–16

Jukka-Pekka S, Pekka A (2003) Reduction of organic emissions by using a multistage drying system for wood-based biomasses. Drying Technol 22(3):541–561

Kenney KL, Cafferty KG, Jacobson JJ, Bonner IJ, Gresham GL, Hess JR, Ovard LP, Smith WA, Thompson DN, Thompson VS, Tumuluru JS, Yancey N (2013) Feedstock supply system design and economics for conversion of lignocellulosic biomass to hydrocarbon fuels. conversion pathway: Biological conversion of sugars to hydrocarbons: The 2017 design case. INL/EXT-13-30342. Idaho National Laboratory, Idaho Falls, ID, USA

Lamers P, Roni MS, Tumuluru JS, Jacobson JJ, Cafferty KG, Hansen JK, Kenney K, Teymouri F, Bals B (2015) Techno-economic analysis of decentralized biomass processing depots. Bioresour Technol 194:205–213

Laurila J, Havimo M, Lauhanen R (2014) Compression drying of energy wood. Fuel Process Technol 124:286–289

Liu Z, Haygreen JG (2007) Drying rates of wood chips during compression drying. Wood Fiber Sci 17:214–227

Nagaya K, Li Y, Jin Z, Fukumuro M, Ando Y, Akaishi A (2006) Low-temperature desiccant-based food drying system with airflow and temperature control. J Food Eng 75(1):71–77

Perinpanayagam M (2013) Mechanical dewatering alternatives evaluation. City of Folsom – Utilities Department. 14 August 2013. Folsom, CA, USA [Online]. Available at: https://www.folsom.ca.us/civicax/filebank/blobdload.aspx?blobid=21076 Accessed 30 April 2020

Roos CJ (2013) Biomass drying and dewatering for clean heat and power. WSUEEP08–015. U.S. Department of Energy CHP Technical Assistance Partnerships, Northwest Division. October 2013. Olympia, WA, USA [Online]. Available at: http://northwestchptap.org/NwChpDocs/BiomassDryingAndDewateringForCleanHeatAndPower.pdf Accessed 30 April 2020

Rupar K, Sanati M (2003) The release of organic compounds during biomass drying depends upon the feedstock and/or altering drying heating medium. Biomass Bioenergy 25(6):615–622

Ståhl M, Berghel J (2008) Validation of a mathematical model by studying the effects of recirculation of drying gases. Drying Technol 26:786–792

Tumuluru JS (2014) Effect of process variables on the density and durability of the pellets made from high-moisture corn stover. Biosyst Eng 119:44–57

Tumuluru JS (2015) High-moisture corn stover pelleting in a flat die pellet mill fitted with a 6 mm die: physical properties and specific energy consumption. Energy Sci Eng 3(4):327–341

Tumuluru JS (2016) Specific energy consumption and quality of wood pellets produced using high-moisture lodgepole pine grind in a flat die pellet mill. Chem Eng Res Des 110:82–97

Tumuluru JS, Yancey N (2018) Conventional and advanced mechanical preprocessing methods for biomass: performance quality attributes and cost analysis. In: Tumuluru JS (ed) Biomass Preprocessing and Pretreatments for Production of Biofuels. CRC Press, Boca Raton, pp 15–49

Tumuluru JS, Cafferty KG, Kenney KL (2014) Techno-economic analysis of conventional, high-moisture pelletization and briquetting process. American Society of Agricultural and Biological Engineers, 2014 Annual Meeting, paper no. 141911360, 13–16 July 2014, Montreal, Quebec

Canada [Online]. Available at: https://elibrary.asabe.org/abstract.asp?aid=45007 Accessed 30 April 2020

Tumuluru JS, Conner CC, Hoover AN (2016) Method to produce durable pellets at lower energy consumption using high-moisture corn stover and a corn starch binder in a flat die pellet mill. J Vis Exp 112:e54092

Tumuluru JS, Yancey N, McCulloch R, Fox C, Conner CC, Hartley D, Dee M, Plummer M (2017) Biomass engineering: size reduction, drying, and densification of high-moisture biomass. Feedstock Supply and Logistics Platform, 2017 Project Peer Review, U.S. Department of Energy (DOE) Bioenergy Technologies Office (BETO), 7 March 2017, Denver, CO, USA [Online]. Available at: https://www.energy.gov/sites/prod/files/2017/05/f34/fsl_tumuluru_1222.pdf Accessed 8 April 2020

Tumuluru JS, Yancey N, Conner CC, Egan E, Dee M, Landon C, Scouten D, Fillerup E, Kane J, Murray DJ (2019) Biomass engineering: size reduction, drying, and densification of high-moisture biomass. Feedstock Supply and Logistics Platform, 2019 Project Peer Review, U.S. Department of Energy (DOE) Biotechnologies Office (BETO), 6 March 2019, Denver, CO, USA [Online]. Available at: https://www.energy.gov/sites/prod/files/2019/04/f61/Size%20Reduction%2C%20Drying%20and%20Densification%20of%20High%20Moisture%20Biomass_NL0026654.pdf Accessed 30 April 2020

U.S. Environmental Protection Agency (EPA) (2000) Biosolids technology fact sheet: centrifuge thickening and dewatering. EPA 832-F-00-053, September 2000. Office of Water, Washington, DC, USA [Online]. Available at: http://www.nesc.wvu.edu/ecommerce/products/WW_GeneralInformation/WWFSGN201DL.pdf

Yancey NA, Tumuluru JS, Wright CT (2013) Drying, grinding, and pelletization studies on raw and formulated biomass feedstocks for bioenergy applications. J Biobased Mater Bioenergy 7(5):549–558

Yoshida T, Sasaki H, Takano T, Sawabe O (2010) Dewatering of high-moisture wood chips by roller compression method. Biomass Bioenergy 34(7):1053–1058

Zlatanović I, Komatina M, Antonijević D (2013) Low-temperature convective drying of apple cubes. Appl Therm Eng 53(1):114–123

Chapter 5
Biochemical and Thermochemical Conversion Performance of Densified Products for Biofuels Production

Abstract The primary question many biofuels producers have is the suitability of densified products for biochemical and thermochemical conversion processes to produce liquid fuels. In general, the belief is the densification of biomass will increase biomass recalcitrance for both biochemical and thermochemical conversions. This chapter will discuss the suitability of densified products such as pellets and briquettes for biochemical conversion and thermochemical conversion pathways, such as gasification, pyrolysis, and combustion.

Keywords Densified products · Pellets · Briquettes · Biochemical conversion · Thermochemical conversion · Biopower

5.1 Introduction

With growing concerns over climate change, there is a significant interest in using biomass for the production of transportation fuels. According to DOE's 2016 Billion-Ton Report, a high potential exists in energy production from biomass. More than one billion tons of agricultural, forestry, waste, and algal biomass currently exist in the United States, which is capable of displacing approximately 30% of petroleum consumption without having any negative impact on the environment, food, or agricultural products (Ray et al. 2017). The major challenge in using biomass for biofuels and biobased product production is variability in moisture, particle size, and density, which creates logistics issues. In addition, other issues related to storage and handling exist. Currently, most biorefineries are having issues related to storage, handling, feeding, and transportation. Typical routes for the production of energy from biomass are biological processes, such as anaerobic digestion and fermentation, thermochemical conversion, such as direct combustion, gasification, pyrolysis, and chemical conversion processes, such as esterification. To efficiently use the biomass resources for both biochemical and thermochemical conversion routes, the supply chain cost, quality of the feedstock, and conversion performance are important as they have an impact on process efficiencies and economics (Ray et al. 2017). According to Tumuluru (2016), large-scale deployment

© Springer Nature Switzerland AG 2020
J. S. Tumuluru, *Biomass Densification*,
https://doi.org/10.1007/978-3-030-62888-8_5

of lignocellulosic biomass is limited by high logistics and preprocessing costs, as well as the low quality of biomass feedstocks.

The inherent physical (e.g., particle size and density) and chemical (e.g., proximate, ultimate, energy properties) characteristics of raw biomass restrict its use for higher percentages in biofuels production. Most of the woody biomass that is typically used for thermochemical conversion is available at high-moisture content values of 30–60% (w.b). Herbaceous biomasses, such as agricultural residues and dedicated energy crops, are available at variable moisture content values ranging from 20 to 35% (w.b.). Biomass needs to be dried to less than a moisture content value of <10% (w.b) and ground to about a 2 mm particle size to improve conversion process efficiencies, especially thermochemical conversion (Bridgewater and Peacocke 2000).

In the case of biochemical conversion, variability in feedstock moisture content is not a major limitation. Still, for thermochemical conversion, both moisture and particle size are critical feedstock specifications. The major challenge in producing particles desirable for the biochemical and thermochemical conversion is variable moisture content, which impacts grinding energy and particle size distribution. Furthermore, grinding raw biomass with high-moisture content is challenging due to its fibrous nature. Tumuluru and Yancey (2018) indicated that grinding energy increases with a corresponding increase in moisture content, which can negatively impact particle size distribution. Other challenges in using high and variable moisture in biomass include the plugging of drop chutes, conveyors, and augurs. All of these challenges result in an uneven flow of biomass into the bioreactor, which affects efficiency. In addition, other problems in terms of chemical composition and CV for commercial-scale production of biofuels and bioproducts also exist. Tumuluru et al. (2016) studied the impact of feedstock supply system unit operations, as well as feedstock quality and cost. These authors concluded that advanced preprocessing and pretreatments help to overcome the significant biomass limitations in terms of physical properties and chemical composition, making feedstock more suitable for both biochemical and thermochemical conversion.

One of the major limitations identified with using biomass for large-scale applications is its low density and variable moisture content. In their review of biomass densification, Tumuluru et al. (2011) concluded that biomass densification is critical to converting biomass feedstock into a commodity-type product. The various advantages with densification include: (a) the production of a product with a definite size and shape that is suitable for the supply chain and biorefinery infeed; (b) an increase by almost four to five times in the bulk density of the biomass, which produces a product with improved controlled particle size distribution and uniformity; (c) fractionated structural components, which can have improved conversion performance; and (d) the development of densified products that meet conversion and supply system specifications. Various densification systems most commonly used for biomass include: (a) the pellet mill; (b) the cuber; (c) the briquette press; (d) the screw extruder; (e) the tablet press; and (f) the agglomerator. Among the various densified products, pellets produced using a pellet mill are commonly used for biofuels and biopower generation. Currently, most biorefineries are not ready to use densified products for biofuels due to pellet production costs. However, the novel grinding and pelleting technologies being developed by industries and national laboratories are helping to reduce pellet production costs. Advanced preprocessing technologies, such as fractional milling, high-moisture pelleting, and low-temperature drying developed by Idaho National Laboratory (INL) have reduced pellet production costs by more than 50% (Tumuluru et al. 2017). This chapter discusses the suitability of the densified products, such as pellets produced using lignocellulosic biomass, for biofuels production.

5.1.1 Biochemical Conversion

In biochemical conversion pathways, biomass is pretreated to give enzymes access to cellulose to convert complex carbohydrates into fermentable sugars. Several chemical pretreatment technologies exist, such as ammonia fiber expansion (AFEX), dilute acid (DA), alkaline, and ionic liquid (IL), to disrupt the crystalline structure of cellulose and depolymerize the lignin. DA pretreatment is a sulfuric acid (0.25–2 wt. %) pretreatment in the presence of high temperature (e.g., 140–200 °C) for a short period ranging from seconds to minutes. This pretreatment helps to solubilize hemicellulose and improves access to cellulose for the enzymes. Alkaline pretreatment helps to solubilize the lignin and deacetylation under relatively mild conditions (e.g., 60–180 °C) with sodium hydroxide (NaOH) or ammonium hydroxide (i.e., soaking in aqueous ammonia [SAA]) as a catalyst (Ray et al. 2017). AFEX pretreatment helps the enzymes to access the cellulose by breaking the crystalline structure of cellulose and depolymerizing the lignin (Hoover et al. 2014). Recently, researchers have also used ionic liquid pretreatment to: (a) break the cell wall and reduce cellulose crystallinity; (b) depolymerize the lignin; and (c) improve the porosity and surface area of the biomass to increase the enzymatic digestibility.

5.1.2 Densification Impact on the Sugar Yields

The physical pretreatment methods commonly used are comminution and densification. Different grinding systems, such as chipping, grinding, and/or milling used for size reduction, can also reduce cellulose crystallinity. Biomass comminution is done in two stages (Yancey et al. 2013). In stage-1, the biomass is ground to bigger screen sizes using 2–6-inch screens and then is further reduced in size during stage-2 with a grinder fitted with smaller screen sizes (e.g., <1-inch (50.4 mm)). According to Kumar et al. (2009), biomass comminution reduces cellulose crystallinity. These authors also indicate the material after chipping has a particle size that is typically somewhere between 10 and 30 mm, while after milling, it generally is between 0.2 and 2 mm. In general, a vibratory ball mill is more effective than an ordinary ball mill in reducing cellulose crystallinity of woody biomass. The grinding energy of the biomass is dependent on the final particle size and moisture content. New grinding technologies are currently developed to make biomass grinding energy less sensitive to moisture and screen size, as indicated by Tumuluru and Yancey (2018). Cadoche and Lopez (1989) reported that if the final particle size is in the range of 3–6 mm, the energy input can be kept below 30 kWh/ton. They also proposed that cellulose irridation by γ-rays leads to cleavage of the alpha 1,4-glycosidic bonds and gives a larger surface area with lower crystallinity. But the challenge becomes whether or not this process is energy-intensive at larger-scale operations.

Low bulk density of ground herbaceous biomass is typically less than 150 kg/m^3, which has a great impact on transportation efficiencies. In addition, low bulk density results in feed handling and storage issues. Biomass is densified using mechanical systems like a pellet mill or a briquette press to overcome this limitation. During densification, biomass is exposed to temperatures of 100 °C and pressures of 4000 psi, which help the biomass to undergo some structural changes that can have a positive impact on conversion performance.

Pellets are commonly used densified products utilized for power generation. Millions of wood pellets produced in the United States and Canada are transported to Europe every year. Currently, great interest exists to use pellets for biochemical conversion to produce sugars and gasification and pyrolysis to produce syngas and bio-oil. The quality of the pellets is influenced by various process variables, feedstock properties, and chemical compositions. During pelleting, the raw material is steam-conditioned and further extruded through a narrow constricted die hole (Tumuluru 2016). Various pelleting process variables, feedstock properties, and feedstock types impact the quality of the densified product produced. Tumuluru (2018) discussed in detail how the various biomass components, such as lignin, cellulose, hemicellulose, protein, starches, and waxes, interact with moisture, temperature, and pressure. During pelleting, the natural binders in the biomass, such as lignin, protein, and starches, get activated and help the particles to agglomerate. Ray et al. (2013) reported that modifications to the biomass components that densification brings could impact its reactivity to pretreatment and enzymatic hydrolysis. Many researchers have tested pellets for biochemical conversion using different pretreatment technologies. Indeed, there is a lot of published literature on how pelleting impacts sugar yields. Most of these studies have indicated that higher sugar release and ethanol yields were observed in a pelleted format when compared to the non-pelleted format after pretreatment and enzymatic hydrolysis.

Guragain et al. (2013) evaluated the effect of alkaline pretreatment on sugar release and ethanol production in pelleted and non-pelleted wheat straw, corn stover, big bluestem, and sorghum stalk. The alkali pretreatment increased the mass recovery by 14%, 11%, 2%, and 5%, respectively, as compared to non-pelleted biomass. Except for sorghum, the volumetric sugar production increased for all the feedstocks. Nahar and Pryor (2014) reported that combining pelleting and pretreatment with SAA treatment reduced cellulase loading to achieve a 90% glucose yield at 10 FPU per gram glucan in switchgrass. Pelleting, the switchgrass did not affect feedstock composition as compared to non-pelleted switchgrass. Bals et al. (2014) tested the susceptibility of AFEX-treated corn stover pellets to enzymatic hydrolysis at high-solids loading (e.g., 18–36%). The studies indicated that pelleting marginally increased the initial rate of hydrolysis relative to raw biomass.

Similarly, Rijal et al. (2012) demonstrated that DA-treated switchgrass did not impact glucose yield in the finer ground and pelleted materials as compared to the native material. However, glucose yields from SAA pretreatment, followed by enzymatic hydrolysis, were higher for both powder and pelleted materials as compared with the native material. Glucose yield for the DA- and SAA-treated and pelleted switchgrass was 98% and 79%, respectively. Theerarattananoon et al. (2012) evaluated the impact of pelleting conditions on sugar release and chemical composition of corn stover, wheat straw, sorghum stalk, and big bluestem grass. DA pretreatment and subsequent enzymatic hydrolysis increased glucan content in the pretreated solids as compared to the non-pelleted companion feedstock for corn stover, wheat straw, and big bluestem prairie grass. Glucan content in pretreated pelleted sorghum stalks was slightly less than that for non-pelleted sorghum stalks. Enzymatic hydrolysis results suggested that pelleting increased cellulose yield for all feedstocks, while wheat straw had the highest cellulose yield at 94.1%.

Dien et al. (2018) studied how pelleted big bluestem, switchgrass, and low-diversity grass mixtures were converted into sugars for bioethanol production. Their studies indicated that pelleting increased bulk density to 528–554 kg/m^3. After pelleting, the glucan and xylan contents reduced slightly during pelletizing (by 23 and 16 g/kg, respectively), as well as theoretical ethanol yields, which are based on total carbohydrate contents. Pretreating pelleted and milled straws with hot water or low-moisture ammonium hydroxide that were further hydrolyzed with cellulases indicated that the glucose and total sugar yields were similar for non-pellets and pellets, but carbohydrates present in the pellets are more efficiently recovered as compared to non-pellets. Figure 5.1 shows the sugar yields from pelletized and non-pelletized samples pretreated with liquid hot-water pretreatment using commercial cellulases and hemicellulases enzyme for hydrolysis (Dien et al. 2018). The sugar yield efficiencies for different feedstocks are also different.

Shi et al. (2013) conducted a series of studies on understanding the impact that mixed feedstocks and feedstock densification have on sugar yields. These authors indicated that mixed feedstocks in a flour or pellet form are processes that efficiently employ the IL pretreatment process. They also indicated that processed four blend feedstocks (i.e., switchgrass, lodgepole pine, corn stover, and eucalyptus) in flour and pellet form using the IL 1-ethyl-3-methylimidazolium acetate could have a positive impact. For both formats, the sugar yields reached 90% within 24 hr. of saccharification. Shi et al. (2013) also used X-ray diffraction (XRD) patterns on the four blend feedstocks, as observed in Fig. 5.2. The cellulose crystallinity index (CrI) for untreated mixed flour and pellets did not change significantly (e.g., 0.37 and 0.33, respectively), suggesting that pelletization did not impact cellulose crystallinity. Treating both the flour and pellets with 1-ethyl-3-methylimidazolium acetate [C2mim][OAc] altered the crystalline structure significantly, and the samples essentially displayed amorphous structures with a limited long-range order, as observed in Fig. 5.2. Calculated residual CrI for the IL-treated mixed flour and pellets are 0.03 and 0.05, respectively (Shi et al. 2013). Figure 5.3 indicates enzyme digestibility (%)

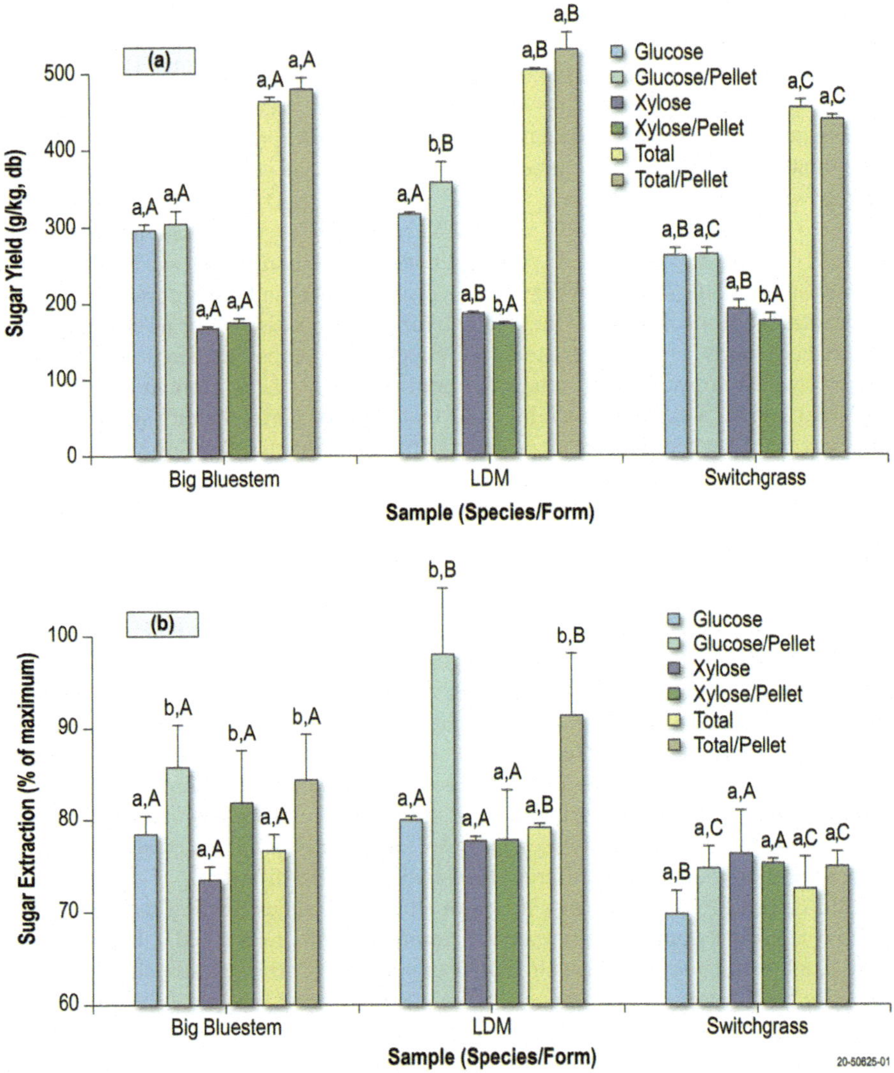

Fig. 5.1 (**a**, **b**) Impact of pelleting on sugar yield and sugar extraction of different biomass feedstocks

with time for treated and untreated flour and pellets made using the four blend feedstock. It is clear that pelleting improves enzyme digestibility as compared with the flour, whereas IL pretreatment significantly improves enzyme digestibility.

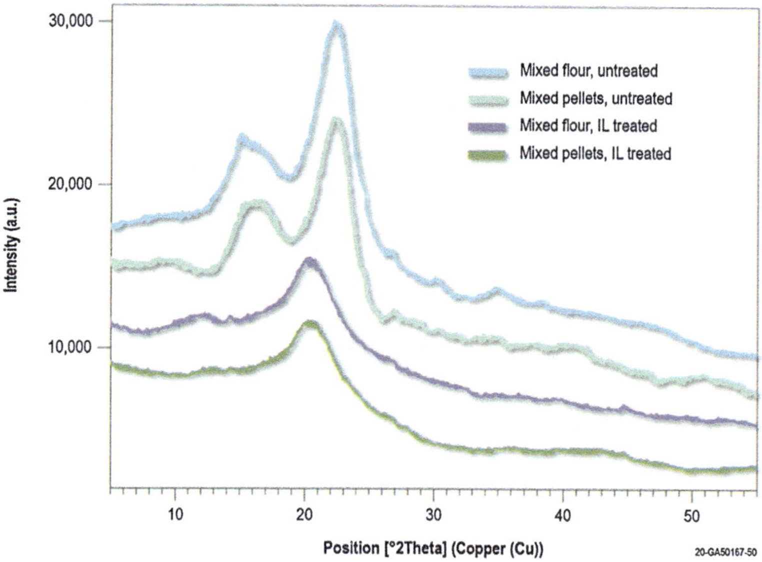

Fig. 5.2 X-ray diffraction patterns of mixed feedstock flour and pellets before and after pretreatment with 1-ethyl-3-methylimidazolium acetate. Pretreatment conditions: biomass loading = 100 g/l, 160 °C for 3 hr. (*IL* ionic liquid)

Ray et al. (2013) studied the effects of pelleting on the recalcitrance and bioconversion of DA pretreated corn stover under low- and high-solids conditions. Pellets were made using corn stover milled to 6.35 mm. These authors also performed low-solids acid pretreatment and simultaneous saccharification and fermentation on both the 2 mm ground and corn stover pelleted using a 6.35 mm hammermill grind. They also studied the sugar yields in two scenarios—low-solids dilute acid pretreatment, and high-solids dilute acid pretreatment. The results indicated that monomeric xylose yields were significantly higher for pellets (~60%) than they were for ground formats (~38%). Pellets achieved approximately 84% of a theoretical ethanol yield; ground stover formats had similar profiles, reaching approximately 68% of theoretical ethanol yield. Figures 5.4 and 5.5 show the xylose and theoretical maximum ethanol yield from low DA pretreatment. Both indicate that the pellets performed better as compared to the products of the 6.35 mm and 2 mm grind. Figure 5.6 shows the percentage of xylose yield from high DA pretreatment, which shows that the pellets produced better xylose and theoretical ethanol yields as compared with those produced from the 6.35 and 2 mm grinds. These authors concluded that pelleting corn stover did not induce any recalcitrance in the biomass, but enhanced ethanol yields. Figure 5.7 indicates the glucose yields for a 6.35 mm grind and pelleted formats using high and low DA solids pretreatment. It is clear from this figure that corn stover in a pelleted format performed better as compared to raw corn stover at lower pretreatment severities.

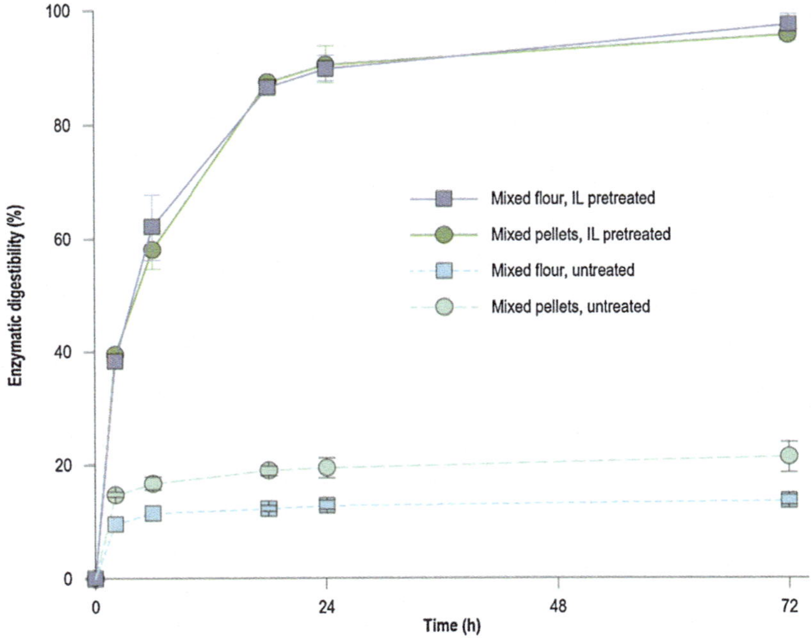

Fig. 5.3 Comparison of enzymatic hydrolysis profiles of untreated and ionic liquid-pretreated mixed feedstock flour and pellets. Biomass loading = 20 g/l; enzyme loading 20 mg CTec2 protein/g glucan and 0.26 mg HTec2 protein/g glucan. (*IL* ionic liquid)

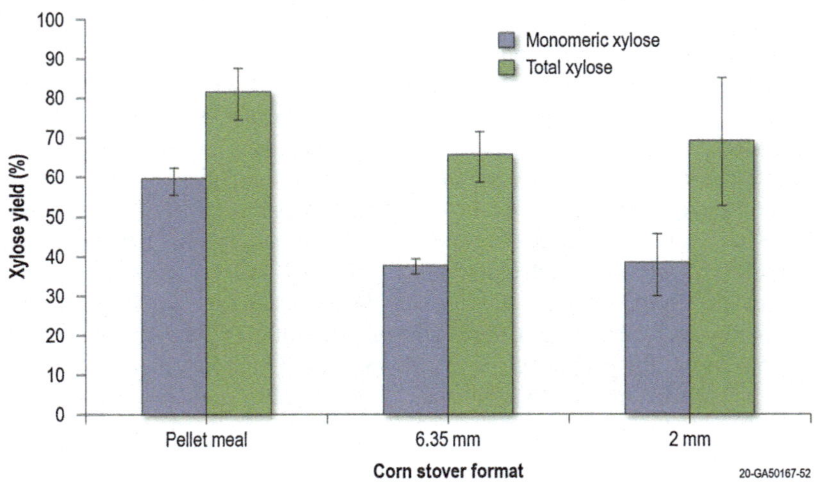

Fig. 5.4 Low-solids, dilute acid pretreatment impact on the monomeric and total xylose yields for different corn stover formats

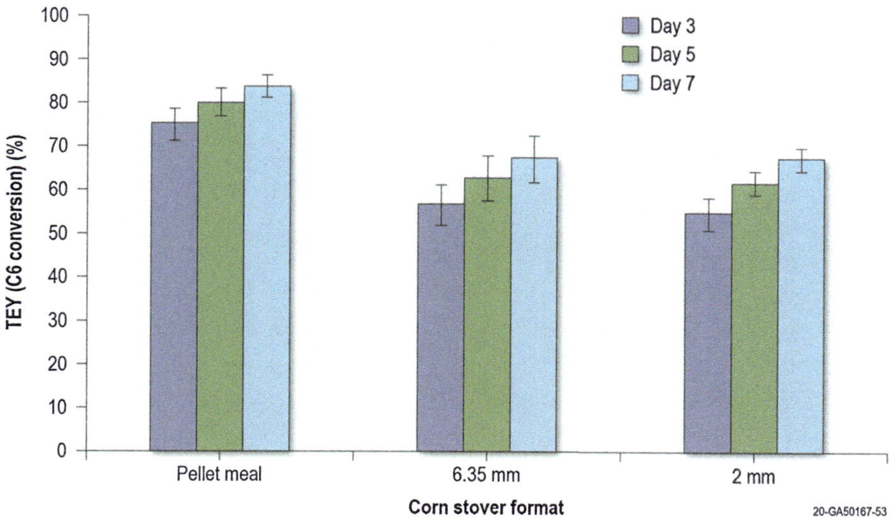

Fig. 5.5 Maximum theoretical ethanol yield for three corn stover formats for 3, 5, and 7 days of simultaneous saccharification and fermentation (note: *TEY* theoretical maximum ethanol yield)

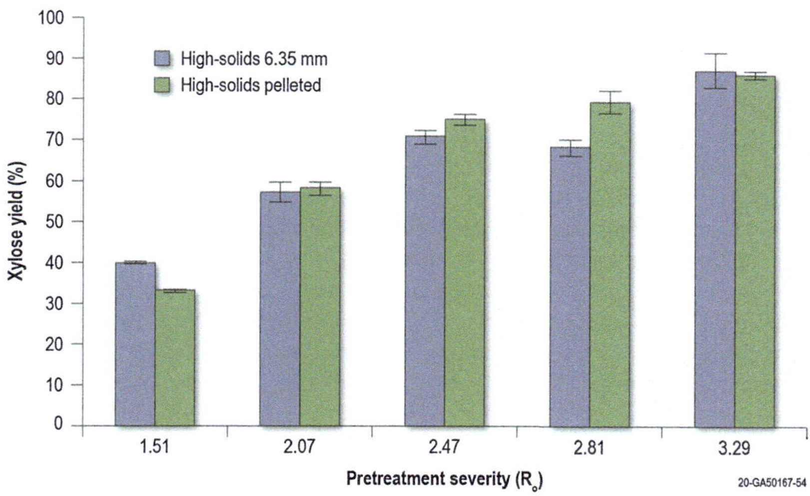

Fig. 5.6 High-solids, dilute acid pretreatment impact on the monomeric and total xylose yields for different corn stover formats

A study conducted by Hoover et al. (2014) indicated that pelleting AFEX-treated corn stover either increased or had no effect on sugar (e.g., glucose and xylose) yields. Pelleting process variables, such as die speed, did not have any impact on the sugar yields for both high- and low-solids loading. Preheating the AFEX-pretreated corn stover before pelleting had either a neutral or a negative effect on sugar yields.

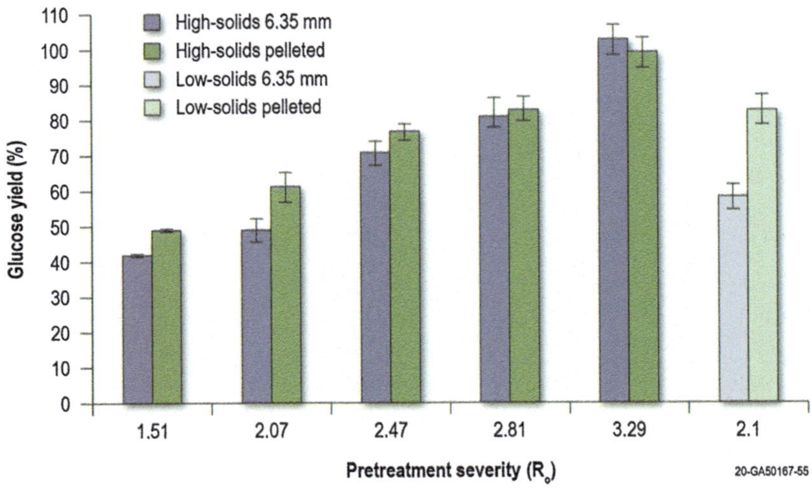

Fig. 5.7 Glucose yields for 6.35 mm and pellets formats using high and low dilute acid solids pretreatment

This research shows that higher pellet die speeds are preferred because they do not negatively impact sugar yields, but can actually increase the throughput of the mill, and in turn, reduce energy consumption. In terms of feedstock properties, a 4 mm particle size is preferred to a 6 mm size due to higher sugar yields and the fact that preheating is not necessary unless it assists other pellet physical properties, such as density, durability, and the overall energy consumption of the process.

Studies conducted by Rijal et al. (2012) corroborates with the findings by Ray et al. (2013) regarding higher sugar yields for pelleted switchgrass. This research clearly indicates that the grinding and pelleting of switchgrass increased the hydrolyzate glucose yields by 37% by SAA. In addition, xylose yields from SAA-pretreated switchgrass pellets were 42% higher than those from the original switchgrass. Figures 5.8 and 5.9 compare the effects of pelleting on xylose and glucose yields (Rijal et al. 2012). Both figures show that pelleting significantly increases the glucose and xylose yields due to its positive effect on the biochemical conversion process. Rijal et al. (2012) concluded that the increase in sugar yields after pelleting could be attributed to grinding and heating of the biomass in a pellet mill during the pelleting process. Rijal et al. (2012) also observed that pelleting alone does not have a negative impact on deacetylation pretreatment efficiency, but improves the SAA pretreatment. They concluded that pelleting could allow the use of less severe pretreatment conditions, or they can reduce enzyme loading and still obtain high sugar yields. In addition to improved yields, pelleting can provide additional supply chain benefits, such as handling, transportation, and storage.

Pandey et al. (2019) studied the pretreatment severity and enzymatic hydrolysis for corn stover pellets following a reduction in SAA. In their studies, pelleted and loose corn stover were treated under the same pretreatment and hydrolysis

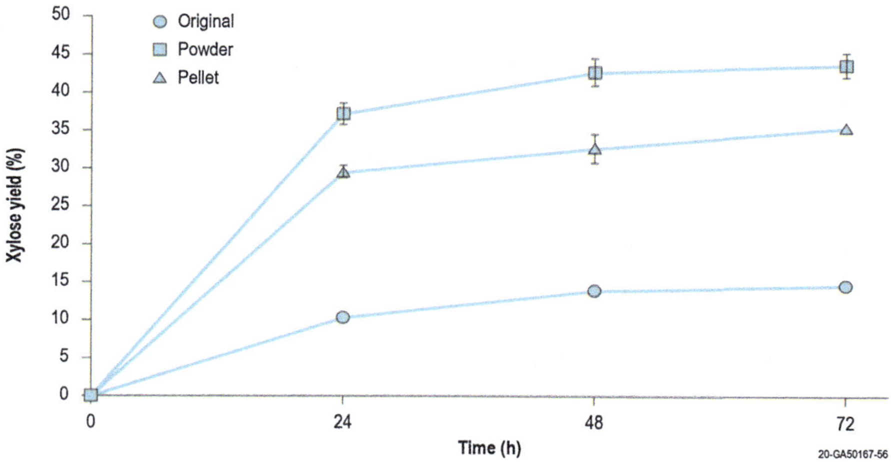

Fig. 5.8 The effect of pelleting on xylose yields

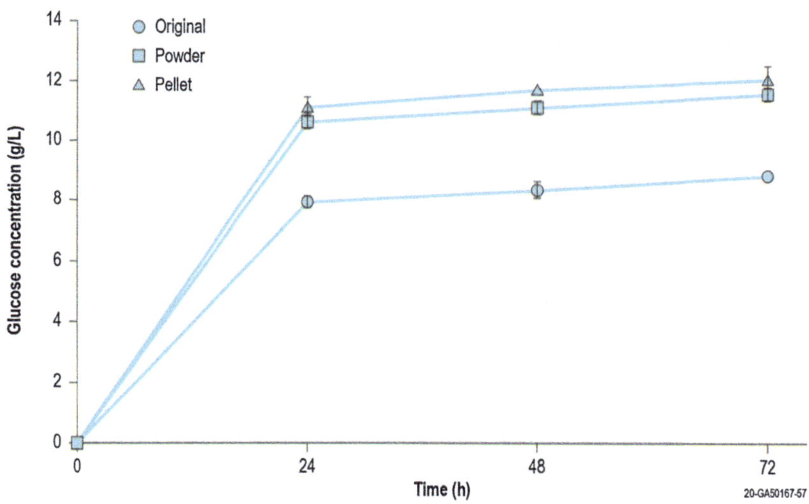

Fig. 5.9 The effect of pelleting on glucose yields

conditions. Table 5.1 indicates the impact of pretreatment conditions on the composition of loose and pelleted corn stover (e.g., solid recovered, glucan, xylan, total lignin, total delignification). Pelleting of switchgrass and higher preheating temperatures, time, and ammonia concentration improved these values. The studies also indicated that glucose yields from pelleted biomass reached 90% with reduced pretreatment severities, enzyme loadings, and hydrolysis time. At the highest enzyme loadings, the use of pelleted corn stover reduced hydrolysis time by up to 58%. Pelleted corn stover also reduced enzyme loading by 80% at higher

Table 5.1 Impact of pretreatment conditions on the composition of loose and pelleted corn stover (derived from Pandey et al. 2019)

Treatment condition	Solid recovered (%)	Glucan (%)	Xylan (%)	Total lignin (%)	Total delignification (%)
Loose corn stover					
Untreated	–	31.2	20.8	18.9	–
PT-1: Temperature: 40 °C, time 4 (h), Ammonia concentration 10 (% as w/v)	84.9	31.1	23.5	17.0	23.6
PT-2: Temperature: 45 °C, time 9 (h), Ammonia concentration 12 (% as w/v)	74.2	35.1	21.9	15.3	40
PT-3: Temperature: 50 °C, time 14 (h), Ammonia concentration 14 (% as w/v)	71.7	40.9	26.5	12.4	52.9
PT-4: Temperature: 55 °C, time 19 (h), Ammonia concentration 16 (% as w/v)	70.5	44.1	27.5	10.8	59.9
PT-5: Temperature: 60 °C, time 24 (h), Ammonia concentration 18 (% as w/v)	65.8	47.8	29.1	10.0	65.3
Pelleted corn stover					
Untreated	–	32	21.0	17.3	–
PT-1: Temperature: 40 °C, time 4 (h), Ammonia concentration 10 (% as w/v)	75.9	39.2	34.1	16.2	28.7
PT-2: Temperature: 45 °C, time 9 (h), Ammonia concentration 12 (% as w/v)	77.3	40.6	25.0	12.7	43.1
PT-3: Temperature: 50 °C, time 14 (h), Ammonia concentration 14 (% as w/v)	72.3	43.3	24.8	11.9	50.4
PT-4: Temperature: 55 °C, time 19 (h), Ammonia concentration 16 (% as w/v)	68.6	47.2	24.3	9.6	50.4
PT-5: Temperature: 60 °C, time 24 (h), Ammonia concentration 18 (% as w/v)	67.0	50.2	25.0	9.7	62.4

pretreatment conditions. At moderate pretreatment conditions, enzyme loading and hydrolysis time was reduced by 40 and 48%, respectively. These authors concluded that the use of pelleted feedstock gives great flexibility to the biorefinery to test different processing options. The results also indicated that the total lignin content decreased after pelleting, which might have resulted in higher solid recovery.

Extrusion pelleting is another densification technology used for biomass. A Karunanithy et al. (2013) study indicated that 28.2, 66.2, and 49.2% of combined sugar recovery could be achieved for switchgrass, big bluestem, and prairie

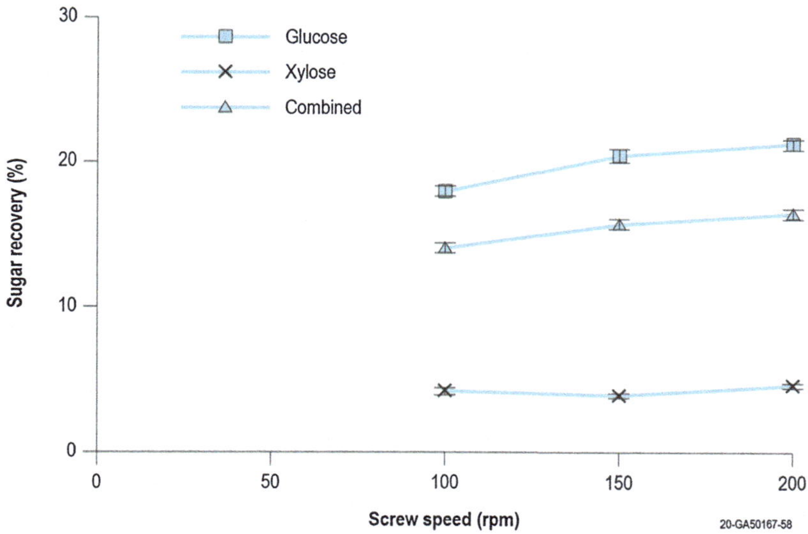

Fig. 5.10 Sugar recovery at different extruder speeds for prairie cordgrass

cordgrass, respectively. The extruder process conditions tested are screw speeds of 200, 200, and 150 rpm, respectively, and barrel temperatures of 75, 150, and 100 °C, respectively, when cellulase and β-glucosidase at a ratio of 1:4 is used. Sundaram and Muthukumarappan (2016) used this ratio to evaluate the effect that laboratory-scale single screw extrusion process variables have on pellet bulk density, hardness, and sugar recovery from enzymatic hydrolysis on AFEX-pretreated corn stover, switchgrass, and prairie cordgrass. The process variables tested included moisture content, hammermill particle size, and extruder barrel temperature. The glucose release from AFEX-pretreated pellets for corn stover was in the range of 88.9–94.9%; for prairie grass, it was about 90.1–94.9%; and for the switchgrass, it was in the range of 87.0–92.9%. The glucose and xylose yields were 1.6, 2.1, 2.3, 1.6, 1.4, and 2.0 fold higher, respectively, for AFEX-treated pellets as compared to untreated pellets. These studies indicate that the extrusion temperature or particle size impacted neither glucose nor xylose. A low extrusion temperature of 75 °C is sufficient to produce pellets desirable for conversion. This research indicated that biomass extrusion does not negatively impact the conversion process, but can positively impact the yields and can also help to reduce the severity of the pretreatment and enzyme loadings, as shown in Fig. 5.10.

Rijal et al. (2014) used a ComPAKco briquetting system to produce briquettes with dimensions of 1 in × 0.5 in × 4 in. The AFEX-pretreated biomass was then subjected to simultaneous saccharification fermentation (SSF) at 4% glucan loading and separate hydrolysis with fermentation (SHF) at 1% glucan loading with an enzyme loading of 15 filter paper units (FPU) and 64 cellobiase units (CBU) per gram of glucan for hydrolysis. The sugar yields of the briquettes made with a 2 mm grind yielded 5% higher ethanol content as compared to the briquettes made with

8 mm AFEX-pretreated materials. The ethanol yields from the SSF did not significantly differ for the AFEX-treated corn stover or switchgrass when compared to the densified AFEX-treated material. Still, they noted a negative impact on the prairie cordgrass densified material by 16%.

Biersbach et al. (2015) studied ethanol yields from AFEX-treated corn stover, switchgrass, and prairie cordgrass briquettes to assess the impact that enzyme loading has during SSF and SHF following long-term storage. Ethanol yields improved for all of the feedstocks after AFEX pretreatment irrespective of enzyme dose or fermentation regimen. These authors concluded that the AFEX-pretreated corn stover briquettes resulted in a 13% increase in ethanol yields. In comparison, the switchgrass and prairie cordgrass briquettes resulted in a 7% and 22% decrease in ethanol yields, respectively. The different ethanol yields for the different feedstocks could be attributed to glucan content, pretreatment efficiency, and enzyme loading.

Steam explosion is a physicochemical pretreatment commonly used for lignocellulosic biomass. In this process, the biomass is subjected to saturated pressure, which is then suddenly released, thereby resulting in explosive decompression. The temperature and pressure are usually in the range of 160–260 °C and 0.69–4.83 MPa, respectively. The time duration of the steam-explosion process varies from a few seconds to a few minutes and is expressed as the severity of the process. This process primarily helps to hydrolyze the hemicellulose, while the lignin transformation increases the cellulose hydrolysis potential. Steam explosion increases the efficiency of enzymatic hydrolysis by 90% as compared to only 15% of the untreated samples, which can be due to the removal of hemicelluloses from the microfibrils and redistribution of lignin and expose the cellulose surface for enzyme accessibility. Researchers are testing novel pretreatment methods to improve sugar yields. Currently, the idea of using supercritical carbon dioxide (CO_2) explosion, which helps to reduce temperature, is being considered as a replacement to steam explosion. CO_2 forms carbonic acid when dissolved in water, which increases the hydrolysis rate of the acid. CO_2 molecules comparable to the size of water and ammonia molecules will help to penetrate small pores that are accessible to water and ammonia molecules. Explosive release of the CO_2 pressure—the disruption of the cellulosic structure—increases the available surface area of the substrate to hydrolysis. Table 5.2 gives a summary of the pretreatment methods used for lignocellulosic biomass.

5.1.3 Thermochemical Conversion

The properties of the various feedstocks after densification were provided in Table 5.2. There are different thermal pretreatment technologies—such as torrefaction, hydrothermal carbonization, and steam explosion—to improve the biomass chemical properties and CV. Integrating densification with these thermal pretreatment methods helps to improve physical properties, such as size, shape, and density.

Table 5.2 Impact of pretreatment on the lignocellulosic biomass (derived from Kumar et al. 2009)

Pretreatment process	Advantages
Mechanical pretreatment	
Comminution	Reduces the cellulose crystallinity
Densification	Presence of heat, pressure, and temperature reduces the cellulose crystallinity
Physicochemical	
Steam explosion	Degrades hemicellulose and distribution of lignin and is cost-effective
Chemical	
Ammonia fiber explosion	(a) Increases the surface area of the biomass (b) Eliminates the lignin and hemicellulose and prevents the formation of inhibitors in the downstream process
CO_2 explosion	(a) Increases the biomass particle surface area (b) More cost-effective (c) Does not cause the formation of inhibitory compounds
Ozonolysis	(a) Reduces lignin content (b) Does not produce toxic residues
Acid hydrolysis	(a) Hydrolyzes hemicellulose to xylose and other sugars (b) Alters lignin structure
Alkaline hydrolysis	(a) Removes hemicellulose and lignin (b) Increases biomass particle surface area
Organosolv	(a) Hydrolyzes lignin (b) Hydrolyzes hemicelluloses
Biological	(a) Degrades lignin and hemicelluloses (b) Low energy

5.1.3.1 Torrefaction

Torrefaction is a thermal pretreatment process where the biomass is typically heated in the range of 200–300 °C for a period of 30 min (Tumuluru et al. 2011). Both the temperature and residence time have an impact on the degree of torrefaction. Torrefaction impacts the physical properties, chemical compositions, and energy properties of the biomass (Tumuluru 2016). Thermal pretreatment makes biomass properties similar to coal, thus making it suitable for cofiring with coal at higher percentages.

Figure 5.11 describes the impact of torrefaction temperature on various biomass components where three temperature regimes are identified: (a) a non-reactive zone that is between 150 and 200 °C; (b) a reactive zone that is between 200 and 250 °C; and (c) a destructive zone that ranges from 250 to 300 °C (Tumuluru et al. 2011). At the non-reactive temperature regime, the biomass loses most of its moisture, which can result in shrinkage. In this temperature zone, the biomass chemical composition remains intact. At the reactive temperature zone, the breakage of hydrogen and carbon bonds will occur and result in the emission of lipophilic extractives. In the destructive temperature zone, most inter- and intra-molecular hydrogen bonds and C-C and C-O bonds are broken, which results in the formation of hydrophilic extractives, carboxylic acids, alcohols, aldehydes, ether, and gases like CO, CO_2,

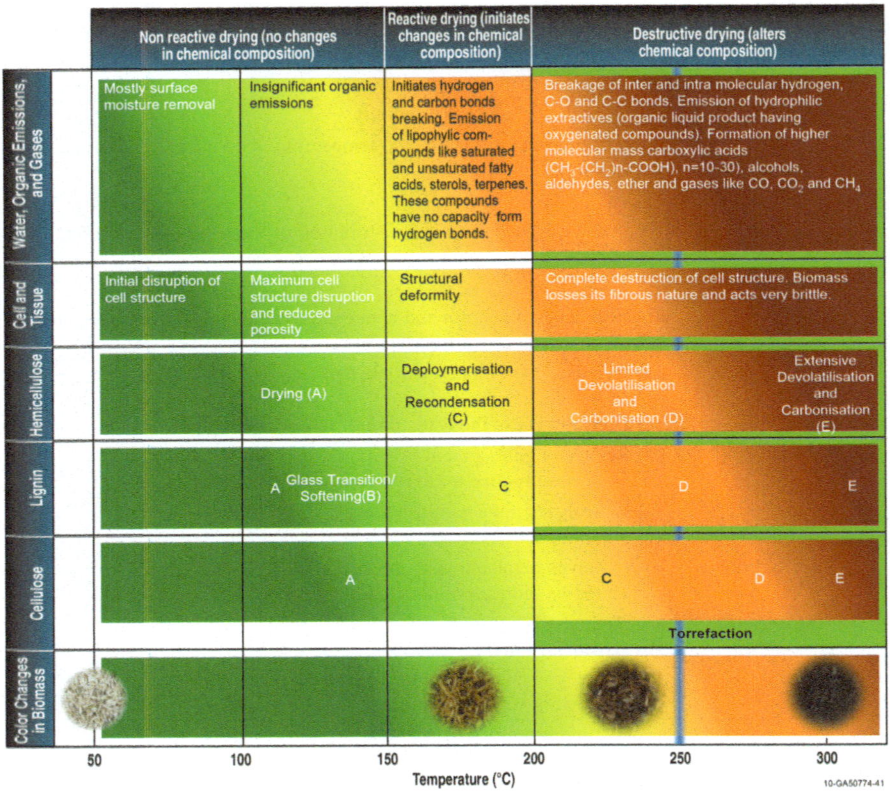

Fig. 5.11 Impact of thermal pretreatment on biomass components (derived from Tumuluru et al. 2011)

and CH_4. At these temperatures, the biomass cell structure is destroyed and also loses its fibrous nature and becomes brittle. In this zone, the typical chemical reactions that take place are dehydration, devolatilization, and carbonization. These reactions result in significant changes in biomass composition and heating values. Figure 5.12 gives higher heating values of raw and torrefied woody biomass samples. Increasing the temperature and residence time increases the higher heating value. The van Krevelen diagram, which is the ratio of H/C and O/C, clearly indicated that torrefied biomass samples are closer to commercially available coals, as observed in Fig. 5.13. Table 5.3 shows the improvements that torrefaction brings in terms of logistics, storage, and physical and chemical properties.

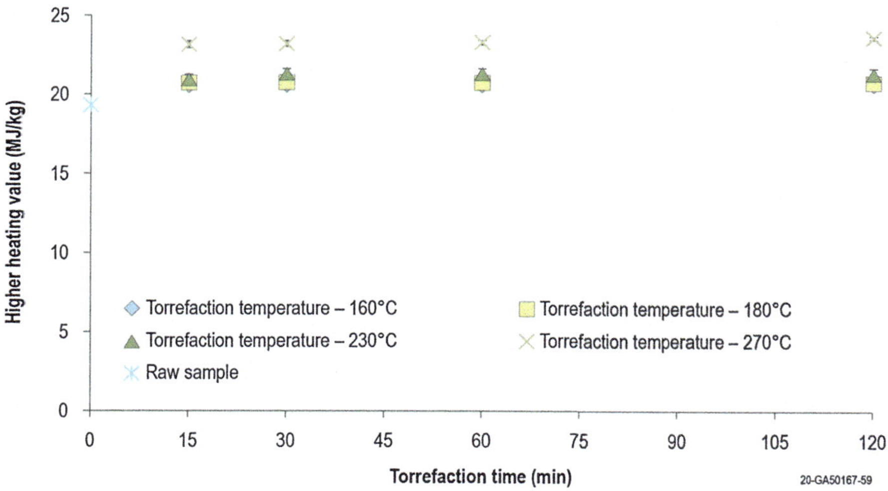

Fig. 5.12 Heating value in lodgepole pine grind at different torrefaction process conditions (derived from Tumuluru 2016)

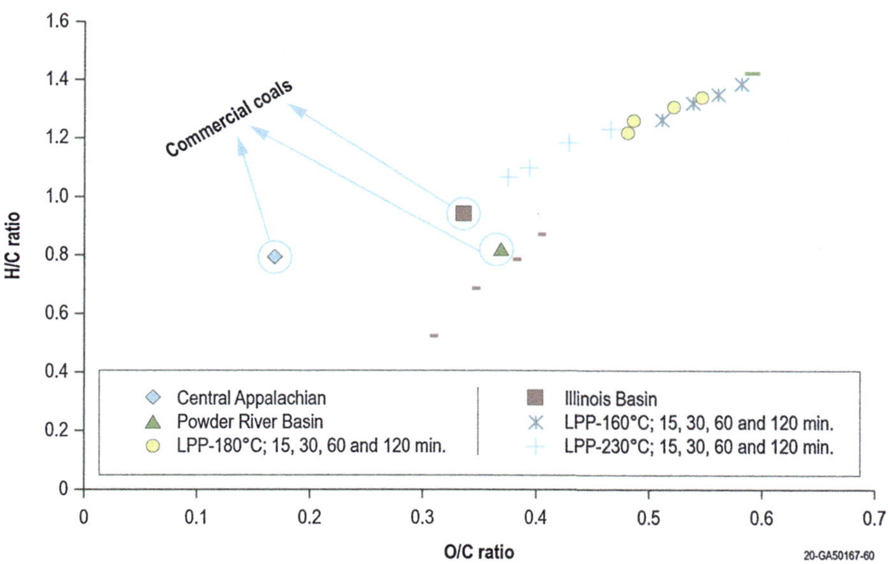

Fig. 5.13 The van Krevelen diagram for raw, torrefied lodgepole pine grind, and commercial coals (derived from Tumuluru 2016)

5.1.3.2 Hydrothermal Carbonization

Hydrothermal carbonization (HTC) converts raw biomass energy-densified hydrochar, which is suitable as a solid biofuel (Hoekman et al. 2018). During the

Table 5.3 Summary of torrefaction effects on biomass quality attributes (derived from Tumuluru and Hess 2015)

RESEARCH FINDINGS

IMPROVED STORABILITY
Hydrophobicity studies (temperature zone D) conducted by Idaho National Laboratory (INL) and Oak Ridge National Laboratory (ORNL) found that torrefaction dramatically reduces the need for expensive covered storage and helps to retain energy value during storage and long-distance transportation.

IMPROVED FLOWABILITY
Grinding studies (temperature zones D and E) at INL have found that torrefied biomass has better particle size and shape sphericity after grinding, which makes it easier to handle in existing high-volume transportation systems and more suitable for thermochemical applications, such as gasification, cofiring and pyrolysis.

IMPROVED ENERGY VALUE
Torrefaction studies (temperature zones D and E) conducted by INL indicated that biomass retains most of its energy content while giving up moisture and low-energy-content volatiles, which increases heating and improves combustion efficiency.

REDUCED LOGISTICS COSTS
The Energy Research Centre of the Netherlands (ECN) reported that torrefaction and densification increase bulk density nearly four-fold and can reduce logistics costs by 30%.

REDUCED PREPROCESSING COSTS
INL grinding studies on deep dired (temperature zone C) and torrefied biomass (temperature zones D and E) found significant reduction in grinding energy required (50-70% less) compared with raw biomass.

REDUCED VARIABILITY
The ECN has found that torrefied biomass (temperature zones D and E) has more consistent moisture content (less than 30%) and pulverises more evenly than untreated biomass, resulting in better blending of varying plant fractions.

REDUCED STORAGE OFF-GASES
INL and University of British Columbia studies indicated torrefied woody biomass (temperature zone D) emits lower (CO, CO_2 and CH_4 off-gases compared with non-torrefied wood chips and commercial wood pellets.

20-GA50167-61

process, the biomass is exposed to hot and pressurized water, usually at temperatures of 200–300 °C and autogenous pressure. The various reactions during hydrothermal carbonization are hydrolysis, dehydration, decarboxylation, condensation-polymerization, and aromatization. During these reactions, the oxygen content in the biomass is removed, and the C/O ratio is increased. The biomass properties, in terms of heating values, grindability, and hydrophobicity, are improved. Also, this process enhances densification characteristics. The degree of properties improvement depends on the severity of the process conditions selected. According to Hoekman et al. (2018), when process severity is low, little energy densification (5–10%) is achieved, whereas, at higher severity, the increase in energy density is about 30–40%. These same authors reviewed the hydrothermal carbonization of various biomass feedstocks at different process conditions. Some of the feedstocks treated with a hydrothermal carbonization process are given in Table 5.4.

5.1.3.3 Steam Explosion

Steam explosion loosens biomass structure and helps to break up large molecules into smaller ones. The severity of deconstruction depends on the severity of the steam treatment (i.e., temperature, the water content of the steam, and treatment duration). According to Lam et al. (2015) and Zimbardi et al. (1999), biomass steam is used as a hydrolysis pretreatment for biochemical conversion. During this process, the biomass is exposed to temperatures between 180 and 240 °C and pressures of 1.03–3.45 MPa for several minutes, which is further depressurized to ambient temperature. The explosive depressurization alters the biomass chemical composition and dimensions in terms of size and shape. According to Shaw et al. (2009), the physiochemical alteration of biomass due to steam explosion results in pellets with a density of 1226 kg/m^3, whereas the untreated pellets had a density of about 1086 kg/m^3 for poplar wood. Studies conducted by Adapa et al. (2010) corroborated with the observations made by Shaw et al. (2009), where steam-exploded wheat straw had small particle size and higher densities after pelleting. Lam (2011) discussed the effect that different steam-explosion temperatures and times have on compression and extrusion energy, as well as the unit density of the produced pellet, as observed in Table 5.5. It is clear that compression and extrusion energy are higher in the steam-treated biomass than is observed in the untreated samples. In addition, there is not much significant difference in solid density. However, steam explosion increases particle bonding and can help to produce pellets with higher durability values. Table 5.6 shows the woody biomass properties of raw, steam-treated, pelletized, and torrefied samples. It is clear from the table that steam-exploded switchgrass pellets had a higher calorific value, density, and durability as compared to the raw switchgrass pellets.

5.1.3.4 Biochemical Conversion of Steam-Exploded Pellets

Tang et al. (2018) studied the influence of the steam-explosion process on the durability and enzymatic digestibility of wood pellets. These authors also tested ground poplar treated with steam at different regimes, including autohydrolysis, SO_2 catalysis, and deacetylation, as well as both acidic and neutral sulfonation for both dried and pelleted poplar. These authors concluded that the use of pelleted feedstock in an enzyme-based biorefinery has a positive impact on the economies of scale. They also accommodate a greater variety of biomass feedstock types for enzyme-based conversion biorefineries. The results of this study indicate that steam explosion improves both the ease of enzymatic conversion and pellet durability at the

Table 5.4 Effect of hydrothermal carbonization process conditions on hydrochar properties (derived from Hoekman et al. 2018)

Biomass	HTC reaction conditions				Hydrochar yield and properties				
Name	Temp (°C)	Hold time	W/B ratio	Severity factor	Mass yield (%)	Energy yield (%)	HHV (MJ/Kg)	Energy density ratio	O/C
Woody biomass									
Loblolly pine	200	1 min	30/1	2.944	79.1	84.8	21.4	1.07	0.62
		3 min		3.421	73.7	79.9	21.7	1.08	0.61
		5 min		3.643	72.9	80.5	22.1	1.10	0.60
Norway spruce	210	30 min		4.716	74.1	80.3	22.4	1.08	0.52
	222	5 min		4.291	73.8	80.7	22.6	1.09	0.51
Loblolly pine	290	20–30 s	3–4	5.293	83.5	101	24.5	1.21	0.48
	290	20–30 s	3–4	5.293	83.5	100	24.3	1.20	0.48
Beech wood chips	210	3 h	7/1	5.494	68	81	22.9	1.19	0.49
Slash pine	235	6 min		4.753	60.9	68.8	22.8	1.13	0.52
	235	12 min		5.054	69.7	80.1	23.2	1.15	0.49
	275	6 min		5.931	54.2	76.4	28.6	1.41	0.26
	275	12 min		6.232	53.6	75.6	28.4	1.41	0.28
Pinyon/Juniper	176	30 min	8/1	3.715	77.4	81.1	21.2	1.05	0.59
	218			5.570	71.5	80.8	23.2	1.13	0.46
	239			6.306	62.7	74.6	24.1	1.19	0.42
	264			6.718	50.6	69.6	28.3	1.38	0.25
	278			7.219	49.6	69.8	28.5	1.41	0.25
	295				48.6	66.3	28.0	1.36	0.23
Herbaceous biomass									
Corn stalk	250	4 h	10/1	6.797	36	60.1	29.2	1.67	0.17
Sugarcane bagasse	180	5 min	10/1	3.054	70.4	75.9	18.5	1.08	0.82
		30 min	10/1	3.833	60.8	66.1	18.6	1.09	0.78

Coconut fiber	220	30 min	10/1	5.010	57	76.7	24.7	1.34	0.37
	250			5.894	45	65.7	26.7	1.45	0.30
	300			7.366	40	65.0	29.4	1.60	0.21
	350			8.838	37	55.8	28.7	1.56	0.21
Corn stover	178	30 min	8/1	3.774	67.0	68.5	18.2	1.02	0.72
	216			4.893	56.6	64.6	20.0	1.14	0.53
	239			5.570	56.4	68.5	21.3	1.21	0.46
	261			6.218	40.0	56.3	24.7	1.41	0.25
	277			6.689	42.4	60.7	25.2	1.43	0.23
	296			7.248	38.0	56.2	26.3	1.48	0.20
Sunflower stalk	190	20 h	20/1	5.729	39.7	58.9	24.5	1.49	
	230			6.907	29.2	50.4	28.5	1.74	
Distillers grains	200	2 h	4/1	5.024	38.9	57.2	29.7	1.47	0.23
Corn stalk	250	4 h	10/1	6.797	36	60.1	29.2	1.67	0.17

Table 5.5 Forces and input energy to make steam-exploded wood pellets and unit density of produced pellets (derived from Lam 2011)

Treatment (steam temperature/ treatment duration)	Max force (N)	Compression Energy[1] (J)	Extrusion Energy[2] (J)	Solid density (g/cm^3)
Untreated	4290	22.3	0.049	1.43
200 °C for 5 min	4190	25.0	0.118	1.42
200 °C for 10 min	4120	31.8	0.117	1.43
220 °C for 5 min	4297	38.8	0.178	1.42
220 °C for 10 min	4079	40.4	0.235	1.42

Table 5.6 Forces and input energy to make steam-exploded wood pellets and unit density of produced pellets (derived from Lam 2011)

Conditions	HHV (MJ/kg)	Density (g/cm^3)	Durability tumbler (%)
Pellets made from steam-treated feedstock and torrefied Douglas fir (DF) pellet (torrefaction at 280 °C for 15 min)	22.8	1.23	99.2
Torrefied Douglas fir (DF) pellet made from untreated biomass (torrefaction at 280 °C for 15 min)	23.2	1.14	98.4
Pellet made from steam-treated Douglas fir (DF) but not torrefied	20.3	1.29	99.7
Douglas fir (DF) pellets made from untreated biomass and not torrefied (control)	18.8	1.17	98.1
Steam-treated and torrefied switchgrass pellet (260 °C for 15 min)	21.3	1.2	99.0
Torrefied switchgrass pellet (260 °C for 15 min)	21.8	1.05	98.2
Steam-treated switchgrass pellet (210 °C for 5 min)	19.0	1.3	99.3
Switchgrass raw pellet	17.3	1.15	97.8

expense of increased energy consumption during the pelletization process. Different pretreatment methods are used to enhance the durability and hydrolyzability of the pellets. Neutral sulfonation during steam pretreatment has helped to recover most of the carbohydrates, increase the hydrophilicity of the lignin, and decrease the enzyme loading for the hydrolysis. The same authors found that increasing the hydrophilicity increases the drying energy; consequently, the durability of the higher heating values becomes lower, thereby making them less suitable for biopower applications. The autocatalysis and SO_2 steam explosion resulted in pellets that are suitable for thermochemical conversion with higher heating value and hydrophobicity. These authors concluded that pretreatment should be integrated into the pellet production process and should be tailored to the end-use application.

Kumar et al. (2012) studied how densification influences the steam-pretreatment and enzymatic hydrolysis of softwood to sugars. These authors indicated that pelleting did not result in significant hemicellulose loss and that steam-pretreated pellets were found to be readily hydrolyzed like steam-pretreated chips. When steam-pretreated prior to pelleting, the wood chips did not require further grinding prior to pelleting. In addition, excellent cellulose hydrolysis was achieved with a

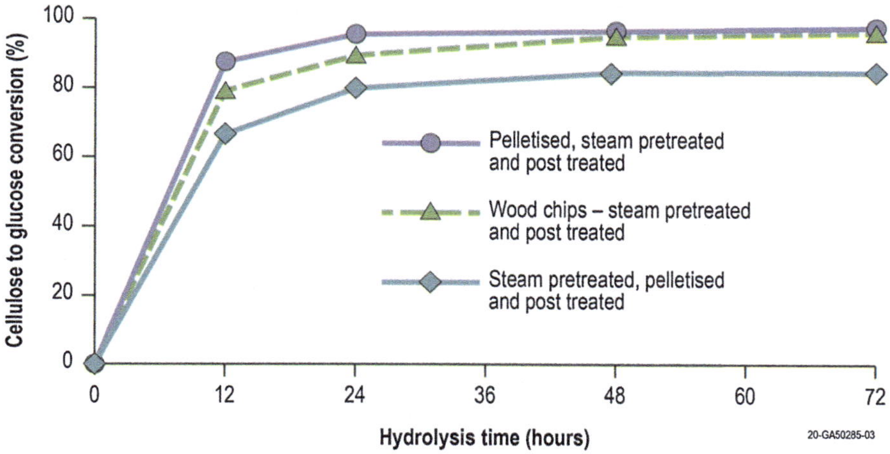

Fig. 5.14 Post-treatment (neutral sulfonation) impact on the enzymatic hydrolysis of the cellulose present in pretreated pellets, wood chips, and steam-stabilized pellets

need for subsequent steam-pretreatment after pelleting. These authors concluded that steam-pretreatment enhanced the pelleting characteristics of chips, good hemicellulose, and cellulose recovery that can be readily hydrolyzed.

Figure 5.14 shows the results of pelleted, steam-treated, and post-treated wood chips reported by Kumar et al. (2012). These authors indicated that sulfonation did not result in significant delignification of the substrates. Both the rate and the extent of enzymatic hydrolysis increased significantly for all the substrates. For the substrate that was steam-pretreated and pelleted, sulfonation was less effective. This study indicated that 85% of the cellulose in the substrate was hydrolyzed, whereas in the case of never-dried and non-pelleted material, more than 90% of the cellulose present in both the steam-pretreated wood chips and pellets was hydrolyzed within 24 hr. (Kumar et al. 2012). It is quite clear from Fig. 5.14 that the hydrolysis rate is slightly higher for the pelleted samples as compared to the wood chips. This slight increase can be due to slight variations in the severity experienced by these substrates during steam-pretreatment and its subsequent influence on sulfonation. Studies by Kumar et al. (2011) indicated that pretreatment severity could bring structural changes to the lignin, which further influences the ability of the post-treatments to bring changes to the lignin (Kumar et al. 2011). Table 5.7 indicates the properties of the densified products prepared using raw and pretreated biomass. It is clear from this table that most of the densified products after thermal pretreatment are closer in composition to coal, indicating that these products can be used as an alternative solid fuel for power generation.

Table 5.7 Properties of raw and pretreated densified biomass compared to coal (derived from Tumuluru et al. 2012)

	Wood chips	Briquettes	Wood pellets	Extruded Logs	FP	SEP	HTP	TOP	Coal
Moisture content (% wt)	30–45	<10	5–7	<8	4–7	5–7	3–6	1–5	10–15
Calorific value (MJ/kg)	9–12	17–19	17–19	19–20	17	20–21	20–25	20–24	23–28
Volatiles (% db)	70–75	70–75	70–75	<60	65	~75–77	70–75	55–65	15–30
Elemental carbon (% bd)	30–35	45	45	45–50	42	53–55	60	55–60	80–90
Bulk density (g/cm^3)	0.2–0.25	0.45–0.55	0.6–0.7	0.3–0.4	0.6–0.7	0.7–0.8	0.75–0.8	0.75–0.8	0.8–0.85
Unit density (g/cm^3)	N/A	0.8–1.0	1.0–1.1	1.3–1.4	1.1–1.2	1.1–1.3	1.1–1.4	1.1–1.4	N/A
Dust	Average	Average	Less	Less	Less	Less	Less	Less	Less
Hygroscopic properties	HPL	HPL	HPL	HPL	HPL	HPB	HPB	HPB	HPB
Biological degradation	Yes	Yes	Yes	Low	Yes	No	No	No	No
Milling requirement	Special	Special	Special	Special	Special	Classic	Classic	Classic	Classic
Handling requirements	Special	Special	Easy	Special	Easy	Easy	Easy	Easy	Easy
Product consistency	Limited	Limited	High	High	High	High	High	High	High
Transport cost	High	Average	Average	Average	Average	Low	Low	Low	Low

Note: *SEP* steam-exploded pellets, *HTTP* hydrothermal pellets, *TOP* torrefied pellets, *HPL* hydrophilic, *HPB* hydrophobic, *FP* (formulated pellets): equal proportions of corn stover, switchgrass, eucalyptus, and pine

5.1.3.5 Thermochemical Conversion

Gasification

Figure 5.15 shows the various biomass reactions that can occur during thermal conversions. Gasification is a process that converts organic or fossil fuels into syngas or synthesis gas or producer gas. This is achieved by subjecting the biomass to a higher temperature of >700 °C under a controlled amount of steam and oxygen. The syngas is a mixture of hydrogen, carbon monoxide, carbon dioxide, methane, and smaller hydrocarbons. The solids that were produced are called char, which are made up of unconverted organic fraction, fixed carbon, and ash. Of all the thermochemical conversion processes, gasification is the most cost-effective process to produce gaseous biofuels. Extensive gasification research has been completed and published in the literature since World War II. There are a series of reactions that occur during gasification in an oxygen-deficient environment—such as drying, pyrolysis, reduction, and oxidation, as observed in Fig. 5.15. Drying temperatures occur at 150 °C, where an increase in temperature causes the carbonaceous materials, such as cellulose and lignin, to become decomposed into small species by cracking the chemical bonds within a temperature range of 250–700 °C. Equation 5.1 provides various products that are produced during biomass gasification. During gasification, there

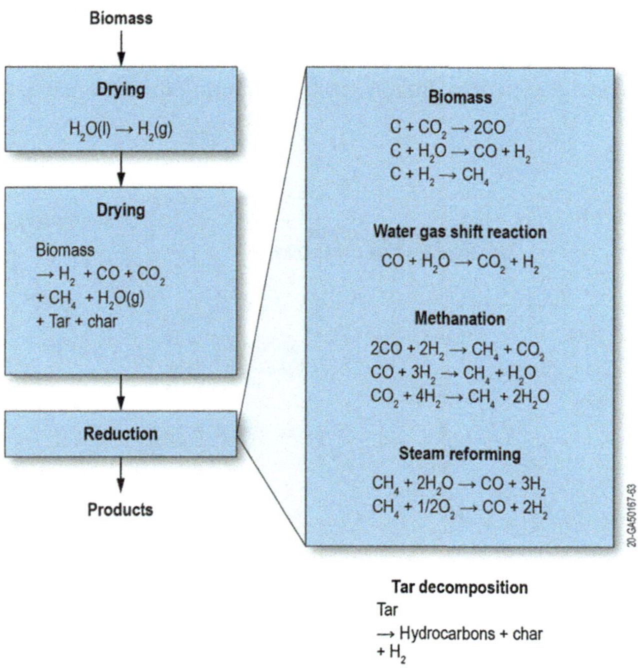

Fig. 5.15 Various reactions during the thermal conversion of biomass (derived from Wang et al. 2017)

are many heat, mass transfer, and reaction kinetics. The reaction temperature impacts the kinetics. At lower temperatures, the heat and mass transfer kinetics are limited. The syngas is formed through the complicated reactions between the gas mixture and the char. The major reactions include the carbonation of char, water gas shift reactions methanation, tar decomposition, and steam reforming.

$$\text{Biomass} \rightarrow H_2 + CO + CO_2 + Hydrocarbons + H_2O(g) + Tar + Char \quad (5.1)$$

Pyrolysis

In the pyrolysis process, the biomass is thermally decomposed in the absence of oxygen. This is achieved during slow heating of biomass over longer residence times of 5–30 min, which increase the production of char products. A fast heating rate with a shorter residence time increases the bio-oil production and valuable chemicals. Decomposition typically occurs relatively at low temperatures between 300 and 600 °C in just a few seconds at atmospheric pressure. Short reaction times and medium temperatures support the production of bio-oil and other chemicals from biomass feedstocks. Figure 5.16 shows the products that are produced during pyrolysis.

Biomass feedstock complexity, pyrolysis, and the production of various products make it challenging to understand the pyrolysis mechanism. During the initial phase, water is removed due to dehydration reactions. During the next stage, the transfer of

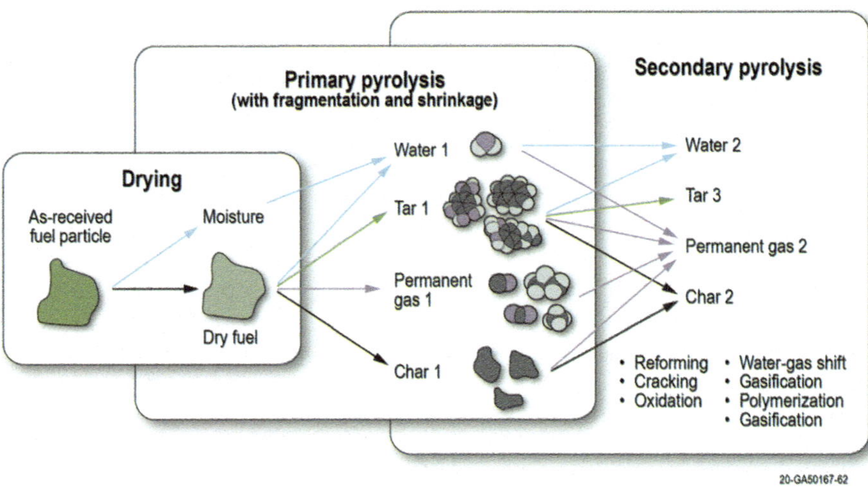

Fig. 5.16 Primary and secondary pyrolysis, including arrows indicating the reaction pathway (thermal conversion of solid biomass particle under inert atmosphere) (derived from Shen et al. 2017)

heat causes the formation of char, bio-oil, gas, and water. Biomass surface properties, such as particle size, shape, and surface area, impact the heat transfer, oil quality, and other products. In general, smaller particle sizes are preferred for pyrolysis. For the pyrolysis process, biomass is typically mechanically ground into a fine powder prior to the reaction. During the secondary stage, further decomposition occurs in addition to various other reactions, resulting in an increase in gaseous products.

Suitability of Densified and Pretreated and Densified Biomass for Gasification and Pyrolysis

Sarkar et al. (2014) studied the gasification performance of pretreated switchgrass, specifically torrefaction and densification. In this study, four switchgrass pretreatments were observed during torrefaction at 230–270 °C, densification, and combined torrefaction and densification at three gasification temperatures (e.g., 700, 800, and 900 °C). The gasification tests were done using a fixed bed externally heated reactor, which uses air as the oxidizing agent. Both the densification and the torrefaction/densification combination had an impact on gasification performance. Results indicated that the pretreatment of switchgrass had a significant effect on syngas yields, the heating value of syngas, carbon conversion, and cold gas efficiencies. Increasing gasification temperature improved the H_2 and CO yields, syngas lower heating value, and gasifier efficiency, whereas the values of CH_4, CO_2, and N_2 were all reduced. Among the different scenarios tested, torrefaction and densification proved to have the best effect for gasification, followed by densified, raw, and torrefied switchgrass, respectively. Table 5.8 indicates the syngas yields from switchgrass for different pretreatment and densification conditions. Gasification of combined torrefied and densified switchgrass resulted in the highest yields of H_2

Table 5.8 Effect of pretreatment on the gas yields for lower heating value, CCE, and CGE

Pretreatment	H_2 yields (kg/kg biomass)	CO yield (kg/kg biomass)	CH_4 yield (kg/kg biomass)	Lower heating value (MJ/m^3)	Carbon conversion effciency (CCE, %)	Cold gas efficiency (CGE, %)
No pretreatment	0.014	0.41	0.10	4.46	87.61	58.28
Torrefaction (230 °C for 30 min)	0.010	0.23	0.11	4.06	67.58	43.43
Torrefaction (270 °C for 30 min)	0.006	0.16	0.12	3.89	58.11	33.46
Densification	0.020	0.49	0.08	4.47	89.58	63.98
Torrefaction (270 °C for 30 min) and densification	0.030	0.63	0.07	4.98	90.69	64.76

Table 5.9 Syngas composition with respect to biomass type

	SP	WP	TWC	PC	MSS
CO (%)	11.9	12.2	7.2	9.0	8.6
CO_2 (%)	9.4	10.0	8.6	8.4	7.2
H_2 (%)	6.53	5.9	8.2	6.2	4.0
CH_4 (%)	2.92	3.10	3.4	1.4	2.8
C_2H_6 (%)	0.4	0.1	0.01	0	na
C_2H_4 (%)	1.23	1.49	0.8	0.01	na
C_3+ (%)	0.14	na	na	0	na
N_2 (%)	67.5	67.2	71.9	75	76.7
H_2/CO	0.87	0.48	1.14	0.69	0.47
Tar (gC/Nm^3)	4.3	9.4	0.99	1.94	32.8
Total S (ppm)	242	22	17	22.3	na
H_2S (ppm)	104	34	12.7	7.5	10
COS (ppm)	19	3.9	1.2	14.5	18
Methylmercaptane (ppm)	0.8	0.4	0.7	na	na

Note: *na* not analyzed, *SP* straw pellets, *WP* softwood pellets, *TWC* torrefied wood chips, *PC* pyrolysis char, *MSS* milled sunflower seed

Table 5.10 Results for biomass gasification experiments

	SP	WP	TWC	PC	MSS
Fluidization velocity (m/s)	0.41	0.43	0.43	0.24	0.17
Residence time (s)	1.22	1.17	1.17	2.11	2.95
Cold gas efficiency (%)	69	56	53	16	17.3
Carbon conversion (%)	82.2	69.7	59.1	25.9	26.7
Gas yield (Nm^3/kg fuel)	2.83	2.47	3.12	2.25	1.31
Calorific value of the syngas per Nm^3 (MJ/Nm^3)	4.40	4.25	3.50	2.33	2.51
Calorific value of the syngas per kggasified fuel (MJ/kg fuel)	12.46	10.52	10.89	5.24	3.29

(0.03 kg/kg biomass) and CO (0.72 kg/kg biomass) and the highest syngas of lower heating value (LHV) (5.08 MJ m^3), carbon conversion efficiency (CCE) (92.53%), and cold gas efficiency (CGE) (68.40%) at the gasification temperature of 900 °C.

Işık-Gulsac et al. (2016) discussed the thermochemical conversion behavior of different biomass feedstocks, such as straw pellet (SP), softwood pellet (WP), torrefied wood chip (TWC), pyrolysis char (PC), milled sunflower seeds (MSS), and distillers grains and solubles (DDGS). On one hand, the results indicate that the highest cold gas efficiency, carbon conversion, and calorific values were obtained for the gasification of SP. On the other hand, SP had some drawbacks regarding its high agglomeration tendency and low deformation temperature. Among all feedstocks, the gasification reactivity of MSS was found to be lower. MSS seemed to expose to pyrolization instead of gasification. WP and TWC were gasified with acceptable conversion values and efficiencies when compared with SP Table 5.9, and Table 5.10 indicates the syngas composition and biomass gasification experiments.

It is clear from the results that WP is the best choice for gasification. Teixeira et al. (2012) studied the gasification of char from wood pellets and wood chips. In their research, these authors used a fixed bed gasifier where char is converted to syngas. The research focus was to understand how the bed compaction impacts the gasification of char by using two types of products: (a) low-density biomass char from wood chips; and (b) high-density char from wood pellets. These two products were gasified in a pilot-scale continuous fixed bed reactor. During gasification, the authors measured the char bed for temperature, gas species concentration, char composition, char bed density, char particle velocity, and specific char and gas sampling techniques. Within the operating conditions, the char bed reactive zone is three times longer for chips (45 cm) than for pellets (16 cm). The results indicate that pelletization has no effect on char bed compaction, final char conversion (about 95%), and syngas quality (16% H_2 and 13% CO). Erlich and Fransson (2011) studied the gasification of pellets made of a wood and palm-oil residue in downdraft gasification. This study has indicated that pellets can be used in a downdraft gasifier. The results show that one and the same reactor can be used for a variety of fuels in pellet form, but at varying air-fuel ratios, temperature levels, gas compositions, and lower heating values. Gasification of wood pellets results in a rich producer gas, while empty fruit bunch (EFB) pellets result in a lower-quality producer gas with higher amounts of non-combustible compounds. In this study, a linear relationship between the air-fuel ratio and cold-gas efficiency was observed, and better efficiencies were observed at higher air-fuel ratios.

Yang et al. (2014a, b) worked on the intermediate pyrolysis of biomass energy pellets to produce sustainable liquid and solid fuels. These authors conducted experiments in pilot-scale systems using energy pellets for pyrolysis. The oil yields were 34.1 wt% and 12.0 wt% for woody and barley straws, respectively. The produced oils are rich in heterocyclic and phenolic compounds and have heating values over 24 MJ/kg. The char produced using woody and barley straw pellets were about 30 wt.% with heating values like sub-bituminous coal. The gas yields were about 20 wt.% with heating values similar to the producer gas produced by using the downdraft gasifier. Based on the produced products, the process efficiency was calculated at about 75%. Table 5.11 gives the various compounds of the bio-oil produced using wood and barley pellets. Interestingly, the study indicated that the wood pyrolysis contained less carbon and higher hydrogen than the barley straws, indicating higher combustion energy. Compositional analysis of bio-oil produced using wood and straw pellets indicated they are similar to that of fossil fuels rich in paraffin, naphthenes, and aromatics. Bio-oil produced from wood pellets had 67.43% of phenolic compounds and 16.94% of furanic acids, which can be due to biopolymers like lignin that is present in the biomasses. For wood oil, aromatic compounds are the major constituents represented by 67.43% of phenolic compounds and 16.94% of furanic compounds. The presence of these compounds was attributable to biopolymer textures of the biomass feedstocks, especially lignin, which is composed of mainly polymerized aromatics.

Studies conducted by Xing et al. (2018) on understanding the characteristics of biochar pellets from corn stover at different pyrolysis temperatures (e.g., 400, 450,

Table 5.11 Compounds of the bio-oil produced using wood and barley straw pellets

Compound	Wood (%)	Barley straw (%)
Furanic	16.94	1.62
Phenolic	67.43	30.68
Aromatic hydrocarbons	–	13.52
Cyclopentenone-based compounds	7.54	8.09
Other heterocyclic	–	7.80
Organic acids	2.93	–
Long-chain hydrocarbon-based compounds (including chloride and nitrile)	–	18.15
Other and unknown	5.16	20.15

500, 550, and 600 °C) indicated that higher heating values increased the yield to 24.55 MJ/kg. These authors concluded that biochar pellets could reduce transportation costs and storage issues. The excellent properties of biochar pellets, in terms of their physical and chemical properties, can help to replace coals, such as lignite and sub-bituminous coal. Pelleting also helped some of the functional groups to weaken or disappear in the biochar showing a compact structure. A Yang et al. (2014a, b) study on the effect of torrefaction and densification on switchgrass pyrolysis products indicated that torrefaction and densification favored the sugars and phenols production. Torrefaction of switchgrass improved the energy and carbon contents. The sugars and phenols are higher for torrefied switchgrass as compared to raw switchgrass. This study also indicated that increasing the torrefaction temperature from 230 °C to 270 °C increased the sugars and phenols content in the pyrolysis products, whereas the guaiacols content decreased. A higher pyrolysis temperature of 600 °C or 700 °C, instead of 500 °C, increased the lignin decomposition, while the anhydrous sugars increased the phenolics, furans, and aromatics. Densification of the torrefied biomass improved the depolymerization of cellulose and hemicellulose during the pyrolysis. Table 5.12 provides a list of the pyrolysis products typically produced at 700 °C for raw switchgrass, densified switchgrass, torrefied switchgrass, and torrefied and densified switchgrass. It is clear from this research conducted by many different researchers that densifying biomass and further using it for biochemical and thermochemical conversion improves fuel quality.

Biopower

The biopower industry is a complex array of power utilities and non-utilities using many technologies and inputs to produce electricity. Biomass is pulverized and cofired with coal. The generated heat converts water into steam. The steam then turns the turbine blades and shaft to produce power. Water is drawn into a condenser, where the steam is turned back into water and reused in the plant. Figure 5.17 shows

Table 5.12 Pyrolysis products of raw, densified, torrefied, and densified switchgrass

Chemicals	Group	Raw switchgrass (6.35 mm screen size)	Pelleted	Torrefaction temperature 230 (°C)	Torrefaction temperature 230 (°C)	Torrefied and pelleted (270 °C + pelleted)
Cellulose/hemicellulose derived compounds (area%)						
Benzofuran, 2,3-dihydro-	Furans	10.71	6.30	3.13	9.69	3.81
Furfural	Furans	3.82	2.94	–	3.17	–
Furan, 2-methyl-	Furans	–	–	2.80	–	2.41
Furan, 2,5-dimethyl-	Furans	–	–	2.23	–	2.12
2-Furancarboxaldehyde, 5-methyl-	Furans	1.71	1.63	1.06	1.17	1.61
2(3H)-Furanone, 5-methyl-	Furans	–	2.46	1.78	1.59	–
5-Hydroxymethylfurfural	Furans	–	1.39	1.21	–	–
2-Furanmethanol	Furans	–	1.22	–	1.33	–
Acetic acid	Acid	4.01	2.07	1.58	–	1.33
d-Allose	Sugars	–	5.78	–	2.82	3.61
β-d-Glucopyranose, 1,6-anhydrous-	Sugars	1.05	6.62	–	4.94	3.55
2-Cyclopenten-1-one, 2-hydroxy-3-methyl-	Ketones	1.08	–	1.30	1.13	1.00
1,2-Cyclopentanedione	Ketones	1.97	1.36	–	–	1.69
Lignin-derived compounds (area%)						
Phenol	Phenols	2.99	2.83	5.01	2.03	3.52
p-Cresol	Phenols	2.95	–	–	–	3.72
Catechol	Phenols	3.90	4.44	3.20	2.80	2.58
Phenol, 2-methyl-	Phenols	1.24	1.02	2.05	2.01	1.43
Phenol, 4-ethyl-	Phenols	1.44	1.46	1.40	–	1.10
Phenol, 2,4-dimethyl-	Phenols	1.19	1.17	1.72	–	1.40

(continued)

Table 5.12 (continued)

Chemicals	Group	Raw switchgrass (6.35 mm screen size)	Pelleted	Torrefaction temperature 230 (°C)	Torrefaction temperature 230 (°C)	Torrefied and pelleted (270 °C + pelleted)
1,2-Benzenediol, 3-methyl-	Phenols	1.64	1.89	–	–	–
1,2-Benzenediol, 4-methyl-	Phenols	2.29	2.76	1.39	1.50	–
2-Methoxy-4-vinylphenol	Guaiacols	4.45	3.79	2.25	6.37	2.35
Creosol	Guaiacols	–	2.57	1.89	1.81	1.08
Phenol, 2-methoxy-	Guaiacols	–	1.47	1.78	1.44	1.28
Phenol, 2,6-dimethoxy-	Syringols	1.37	2.07	1.72	2.04	1.18
Toluene	Aromatics	3.80	3.15	4.30	2.97	3.86
Benzene	Aromatics	2.55	1.61	3.49	–	3.07
Xylene	Aromatics	1.24	1.16	1.42	–	–
Styrene	Aromatics	1.28	–	1.78	–	1.22

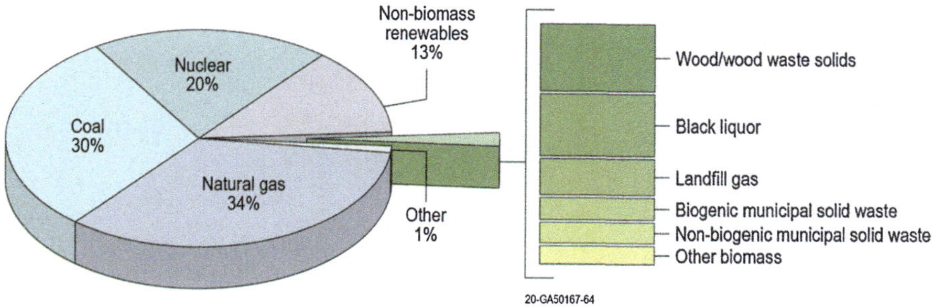

Fig. 5.17 Power generation in the U.S. using various fuel sources (EIA 2017)

U.S. electricity generation by fuel type in 2016. Biomass and waste fuels made up about 2% of this total.

In 2016, biomass and waste fuels generated about 71.4 billion kilowatt-hours (U.S. Energy Information Agency [EIA], 2017). Biomass is defined as any non-fossil, carbon-based energy source, whereas waste is defined as non-biogenic waste. Wood-based solid biomass, such as chips and other densified products, accounted for nearly one-third of the electricity generated from biomass and waste. The major sources of woody biomass are logging and mill residues, wood, paper, discarded large timber products, furniture manufacturing waste, railway ties, utility poles, and marine pilings (EIA 2017). Wood-derived fuels that are by-products from the chemical processing of wood are also used for biofuels production. Wood-derived fuels are the by-product produced during the making of paper-related products. The black liquor is the major wood-derived product made from the paper-making process. In 2016, black liquor accounted for 27% of the electricity generated from waste. Other waste products derived from the paper-making process include sludge waste, wood-waste liquids, and other biomass liquids, which contributed to about 0.5% of the electricity. Another major resource of renewable energy from biobased material is municipal solid waste (MSW). EIA estimates that biogenic sources such as wood, paper, food, rubber, and yard trimmings contributed to about 51% of MSW-based electricity in 2016 (EIA 2017).

Wood Pellets for Biopower Generation

The United States exports most of its wood pellets produced to Europe. According to data released by the U.S. Department of Agriculture (USDA), 546,435.2 metric tons of wood pellets were exported by the United States in February 2020, which is higher by 500,904 tons in January 2020 and 501,563.9 tons in February 2019 (Voegele 2020). These pellets were exported to 18 countries. About 418,529.9 tons of pellets were exported to the United Kingdom, 91,970.5 tons to Belgium-Luxembourg, and 32,137.5 tons to the Netherlands. The value of U.S. wood pellet exports reached $76.51 million in February 2020, up from $56.95 million in January

Fig. 5.18 Pellets used for power generation

Fig. 5.19 Power generation by burning the pellets

2020, and $63.73 million in February 2019. During the first two months, the United States exported nearly 1.05 million tons of wood pellets at a value of $145.82 million, whereas it is about 888,342.7 tons at a value of $117.65 million during the same period of 2019 (Voegele 2020). In the United States, Drax Corporation established pellet production facilities that have been running for the last five years. The wood pellets produced from this company were transported from Baton Rouge, LA, USA, to the United Kingdom's largest power station, which was then used to generate electricity for millions of homes in the United Kingdom. Drax has three pellet mills in the United States which use the wood from the forests of Louisiana, Arkansas, and Mississippi to produce wood pellets. About two-thirds of the 7.5 million metric tons of wood pellets that Drax uses each year comes from the United States. Figures 5.18 and 5.19 show how the pellets can be used in boilers for biopower generation.

5.2 Conclusions

Currently, biorefineries are not operating at designed capacities due to storage, transportation, and handling issues. The preprocessing of biomass helps to not only improve the logistics, storage, and feed handling processes, but biomass conversion efficiencies as well. The challenge in preprocessing is the cost, which adds additional costs to feedstock preparation overall. However, the advantages it provides in terms of reducing storage and transportation costs by improving feed handling and conveyance are not quantified in terms of cost, as it has never been studied in detail. Improving transportation and storage logistics, as well as improving feed handling and conveyance, provides an opportunity for biorefineries to operate at designed capacities and thereby increase their profitability. Studies indicated that biomass preprocessing not only improves feedstock characteristics but also improves conversion efficiency. They also indicate that pelleted biomass improves the biochemical conversion of herbaceous biomass and produces higher sugar yields. The thermochemical conversion of pelleted biomass also indicated that the quality of syngas and bio-oil yields are better as compared to the raw biomass tested. The physical and mild chemical changes produced in the biomass during densification is one of the primary reasons for better performance during conversion.

References

Adapa PK, Tabil LG Jr, Schoenau GJ (2010) Compression characteristics of non-treated and steam exploded barley, canola, oat, and wheat straw grinds. Appl Eng Agric 26(4):617–632

Bals BD, Gunawan C, Moore J, Teymouri F, Dale BE (2014) Enzymatic hydrolysis of pelletized AFEX™-treated corn stover at high solid loadings. Biotechnol Bioeng 111(2):264–271

Biersbach G, Rijad B, Pryor SW, Gibbons WR (2015) Effects of enzyme loading, densification, and storage on AFEX-pretreated biomass for ethanol production. Appl Biochem Biotechnol 177 (7):1530–1540

Bridgewater AV, Peacocke GVC (2000) Fast pyrolysis processes for biomass. Renew Sustain Energy Rev 4:1):1–1)73

Cadoche L, Lopez GD (1989) Assessment of size reduction as a preliminary step in the production of ethanol from lignocellulosic wastes. Biol Wastes 30:153–157

Dien BS, Mitchell RB, Bowman MJ, Jin VL, Quarterman J, Schmer MR, Singh V, Slininger PJ (2018) Bioconversion of pelletized big bluestem, switchgrass, and low-diversity grass mixtures into sugars and bioethanol. Front Energy 6:1–17

EIA (2017) Biomass and waste fuels made up 2% of total US electricity generation in 2016 [Online]. Available at: https://www.eia.gov/todayinenergy/detail.php?id=33872#:~:text=Biomass%20and%20waste%20fuels%20generated,based%20(biogenic)%20energy%20sources) Accessed 23 June 2020

Erlich C, Fransson TH (2011) Downdraft gasification of pellets made of wood, palm-oil residues respective bagasse: experimental study. App Eng 88(3):899–908

Guragain YN, Wilson J, Staggenborg S, McKinney L, Wang D, Vadlani PV (2013) Evaluation of pelleting as a preprocessing step for effective biomass deconstruction and fermentation. Biochem Eng J 77:198–207

Hoekman SK, Broach A, Felix L (2018) Hydrothermal carbonization (HTC) of biomass for energy applications. In: Tumuluru JS (ed) Biomass Preprocessing and Pretreatments for Production of Biofuels. CRC Press, Boca Raton, pp 196–254
Hoover AN, Tumuluru JS, Teymouri F, Moore J, Gresham G (2014) Effect of pelleting process variables on physical properties and sugar yields of ammonia fiber expansion pretreated corn stover. Bioresour Technol 164:128–135
Işık-Gulsac I, Durak-Cetin Y, Engin B, Gafarova-Aksoy P, Karataş H, Sarıoğlan A (2016) Thermochemical conversion behavior of different biomass feedstocks: pyrolysis and gasification. J Turkish Chem Soc A 3(3):731–746
Karunanithy C, Muthukumarappan K, Gibbons W (2013) Effect of extruder screw speed, temperature, and enzyme levels on sugar recovery from different biomasses. ISRN Biotechnol 2013:942810
Kumar P, Barrett DM, Delwiche MJ, Stroeve P (2009) Methods for pretreatment of lignocellulosic biomass for efficient hydrolysis and biofuels production. Ind Eng Chem Res 48:3713–3729
Kumar L, Chandra R, Saddler J (2011) Influence of steam-pretreatment severity on post-treatments used to enhance the enzymatic hydrolysis of pretreated softwoods at low enzyme loadings. Biotechnol Bioeng 108:2300–2311
Kumar L, Tooyserkani Z, Sokhansanj S, Saddler JN (2012) Does densification influence the steam-pretreatment and enzymatic hydrolysis of softwoods to sugars? Bioresour Technol 121:190–198
Lam PS (2011) Steam-explosion to biomass to produce durable wood pellets. Ph.D. Thesis, University of British Columbia, Vancouver, British Columbia, Canada
Lam PS, Lam PY, Sokhansanj S, Bi XT, Lim CJ, Melin S (2015) Effect of steam-explosion pretreatment on size reduction and pellet quality of woody and agricultural biomass. In: Tannous K (ed) Innovative Solutions in Fluid-Particle Systems and Renewable Energy Management. IGI Global, Hershey, pp 27–53
Nahar N, Pryor SW (2014) Reduced pretreatment severity and enzyme loading enabled through switchgrass pelleting. Biomass Bioenergy 67:46–52
Pandey R, Nahar N, Tumuluru JS, Pryor SW (2019) Quantifying reductions in soaking in aqueous ammonia pretreatment severity and enzymatic hydrolysis conditions for corn stover pellets. Bioresour Technol Rep 7:100187
Ray AE, Hoover AN, Nagle N, Chen X, Gresham G (2013) Effect of pelleting on the recalcitrance and bioconversion of dilute-acid pretreated corn stover under low- and high-solids conditions. Biofuels 4(3):271–284
Ray AE, Li C, Thompson VS, Daubaras DL, Nagle NJ, Hartley DS (2017) Biomass blending and densification: Impacts on feedstock supply and biochemical conversion performance. In: Tumuluru JS (ed.) Biomass Volume Estimation and Valorization for Energy. IntechOpen. https://doi.org/10.5772/67207
Rijal B, Igathinathane C, Karki B, Yu M, Pryor SW (2012) Combined effect of pelleting and pretreatment on enzymatic hydrolysis of switchgrass. Bioresour Technol 116:36–41
Rijal B, Biersbach G, Gibbons WR, Pryor SW (2014) Effect of initial particle size and densification on AFEX-pretreated biomass for ethanol production. Appl Biochem Biotechnol 174 (2):845–854
Sarkar M, Kumar A, Tumuluru JS, Patil KN, Bellmer DD (2014) Gasification performance of switchgrass pretreated with torrefaction and densification. Appl Energy 127:194–201
Shaw MD, Karunakaran C, Tabil LG Jr (2009) Physicochemical characteristics of densified untreated and steam-exploded poplar wood and wheat straw grinds. Biosyst Eng 103 (2):198–207
Shen Y, Ma D, Ge X (2017) Co2-looping in biomass pyrolysis or gasification. Sustainable Energy & Fuels 1:1700–1729
Shi J, Thompson VS, Yancey NA, Stavila V, Simmons BA, Singh S (2013) Impact of mixed feedstocks and feedstock densification on ionic liquid pretreatment efficiency. Biofuels 4 (1):63–72

Sundaram V, Muthukumarappan K (2016) Impact of AFEX™ pretreatment and extrusion pelleting on pellet physical properties and sugar recovery from corn stover, prairie cord grass, and switchgrass. Appl Biochem Biotechnol 179(2):202–219

Tang Y, Chandra RP, Sokhansanj S, Saddler JN (2018) Influence of steam-explosion processes on the durability and enzymatic digestibility of wood pellets. Fuel 211:87–94

Teixeira G, Van de Steene L, Martin E, Gelix F, Salvador S (2012) Gasification of char from wood pellets and from wood chips: textural properties and thermochemical conversion along a continuous fixed bed. Fuel 102:514–524

Theerarattananoon K, Xu F, Wilson J, Staggenborg S, McKinney L, Vadlani P, Pei XJ, Wang DH (2012) Effects of the pelleting conditions on chemical composition and sugar yield of corn stover, big bluestem, wheat straw, and sorghum stalk pellets. Bioprocess Biosyst Eng 35 (4):615–623

Tumuluru JS (2016) Specific energy consumption and quality of wood pellets produced using high moisture lodgepole pine grind in a flat die pellet mill. Chem Eng Res Des 110:82–97

Tumuluru JS (2018) Thermal pretreatment of biomass to make it suitable for biopower application. In: Tumuluru JS (ed) Biomass Preprocessing and Pretreatments for Production of Biofuels. CRC Press, Boca Raton, pp 255–291

Tumuluru JS, Hess JR (2015) New market potential: torrefaction of woody biomass. Materials World [Online]. Available at: https://www.iom3.org/materials-world-magazine/feature/2015/jun/02/new-market-potential-torrefaction-woody-biomass Accessed 3 June 2020

Tumuluru JS, Yancey N (2018) Conventional and advanced mechanical preprocessing methods for biomass: performance quality attributes and cost analysis. In: Tumuluru JS (ed) Biomass Preprocessing and Pretreatments for Production of Biofuels. CRC Press, Boca Raton, pp 15–49

Tumuluru JS, Wright CT, Hess JR, Kenney KL (2011) A review of biomass densification systems to develop uniform feedstock commodities for bioenergy application. Biofuels Bioprod Biorefin 5(6):683–707

Tumuluru JS, Hess JR, Boardman RD, Wright CT, Westover TL (2012) Formulation, pretreatment, and densification options to improve biomass specifications for cofiring high percentages with coal. Ind Biotechnol 8(3):113–132

Tumuluru JS, Searcy E, Kenney KL, Smith WA, Gresham GL, Yancey NA (2016) Impact of feedstock supply systems unit operations on feedstock cost and quality for bioenergy applications. In: Kumar R, Singh S, Balan V (eds) Valorization of Lignocellulosic Biomass in a Biorefinery: From Logistic to Environmental and Performance Impact. Nova Science Publishers, Inc., Hauppauge, pp 1–45

Tumuluru JS, Yancey N, McCulloch R, Fox C, Conner CC, Hartley D, Dee M, Plummer M (2017) Biomass engineering: Size reduction, drying, and densification of high moisture biomass. Feedstock Supply and Logistics Platform, 2017 Project Peer Review, US Department of Energy (DOE) Bioenergy Technologies Office (BETO), 7 March, Denver, CO, USA [Online]. Available at: https://www.energy.gov/sites/prod/files/2017/05/f34/fsl_tumuluru_1222.pdf Accessed 8 April 2020

Voegele E (2020) USDA: US wood pellet exports top 546,435 tons in February. Biomass Magazine [Online]. Available at: http://www.biomassmagazine.com/articles/16940/usda-us-wood-pellet-exports-top-546-435-tons-in-february Accessed 15 June 2020

Wang AG, Austin D, Song H (2017) Catalytic biomass valorization. In: Tumuluru JS (ed.) Biomass Volume Estimation and Valorization for Energy. IntechOpen. https://doi.org/10.5772/65826

Xing X, Fan F, Jiang W (2018) Characteristics of biochar pellets from corn straw under different pyrolysis temperatures. R Soc Open Sci 5(8):172346

Yancey NA, Tumuluru JS, Wright CT (2013) Drying, grinding, and pelletization studies on raw and formulated biomass feedstocks for bioenergy applications. J Biobased Mater Bioenergy 7 (5):549–558

Yang Y, Brammer JG, Mahmood ASN, Hornung A (2014a) Intermediate pyrolysis of biomass energy pellets for producing sustainable liquid, gaseous, and solid fuels. Bioresour Technol 169:794–799

Yang Z, Sarkar M, Kumar A, Tumuluru JS, Huhnke RL (2014b) Effects of torrefaction and densification on switchgrass pyrolysis products. Bioresour Technol 174:266–273
Zimbardi F, Viggiano D, Nanna F, Demichele M, Cuna D, Cardinale G (1999) Steam-explosion of straw in batch and continuous systems. Appl Biochem Biotechnol 77–79:117–125

Chapter 6
International Standards for Densified Products

Abstract The primary solid fuel standard used in the United States is the Pellet Fuel Institute (PFI) standard. In contrast, combined standards throughout various European countries—such as Austria, Sweden, the United Kingdom, France, and Denmark—are governed by what is known as the European Committee for Standardization (CEN). CEN/TC 335 developed technical specifications and testing methods for solid biofuels. CEN/TS 14961 gives the standards for densified solid fuels like pellets and briquettes. International standards for pellets and briquettes (ISO 17225-1:2014) are guided by the International Standards Organization (ISO). This chapter compares the quality of these densified products produced from various woody, herbaceous, and municipal solid waste (MSW) with respect to the existing standards.

Keywords Pellet quality · Physical properties · Chemical composition · International standards · Pellet Fuel Institute (PFI) · European Committee for Standardization (CEN) · International Standards Organization (ISO)

6.1 Introduction

Biomass pellets are considered a renewable, clean-burning fuel. In general, recycled wood waste is used for making pellets. There is also quite a lot of other woody and herbaceous biomass sources, such as cordwood, wood pellets, wood chips, and waste paper, along with dozens of other agricultural products and by-products to produce pellets for energy applications. The pellets are preferred densified products as they are consistent in terms of size, shape, and density, as well as their burn efficiency and their reduced particulate emissions. In addition, biomass has become sustainable with the management practices that are currently developed. The significant advantages of using pellets for power generation are: (a) low moisture; (b) higher density; (c) uniform size and shape, which help to feed smaller systems; and (d) easier handling and improved transportation efficiencies.

Pellet production has grown significantly throughout the world in the last 20 years. There are many factors that have contributed to this accelerated growth.

© Springer Nature Switzerland AG 2020

J. S. Tumuluru, *Biomass Densification*,
https://doi.org/10.1007/978-3-030-62888-8_6

These factors include: (a) the use of pellets in coal power plants; and (b) the increased use of pellets in pellet stoves and boilers. Europe has significantly increased the production of pellets to the point that it is now responsible for producing and supplying up to 51% of the market share (Cocchi et al. 2011). Germany, Sweden, Latvia, and Austria are the primary countries in Europe that are contributing to the vast majority of this pellet production (Cocchi et al. 2011).

North America provides a substantial amount of wood pellet production as well. The United States contributes about 27% of the world's total wood pellet production, while Canada is responsible for another 7%. Most of the pellets produced in the United States and Canada are often shipped overseas for use in Europe. The European market is the primary reason for the expansion of the wood pellet industry in Canada, especially in British Columbia. Among the 1.3 million tons of pellets exported from Canada in 2007, 600,000 tons were exported to Europe; the main contributor of this supply is British Columbia (Will 2009). With vast forest resources in North America, the potential exists for a great production. Indeed, many companies are looking to tap into this expanding market. Currently, a huge demand for wood pellets exists in Asia, as many countries are planning to phase out coal power plants.

Figure 6.1 shows the world's overall wood pellet production shares as they stood in 2016. The figure clearly indicates that Europe provides 39% of this production, followed by North America at 27%, and China providing another 20%. Figure 6.2 indicates the global pellet outlook up to 2024. It is also evident from the figure that the use of these pellets in both industrial and heating applications has significantly grown since 2004 and is expected to continue this trend over the next 5 years.

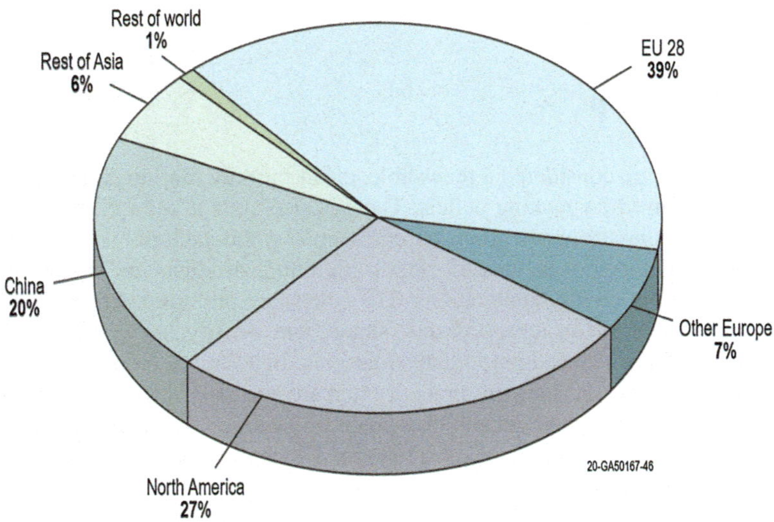

Fig. 6.1 Percentage of world wood pellet production shares in 2016 (derived from the International Energy Agency EPC Survey 2017)

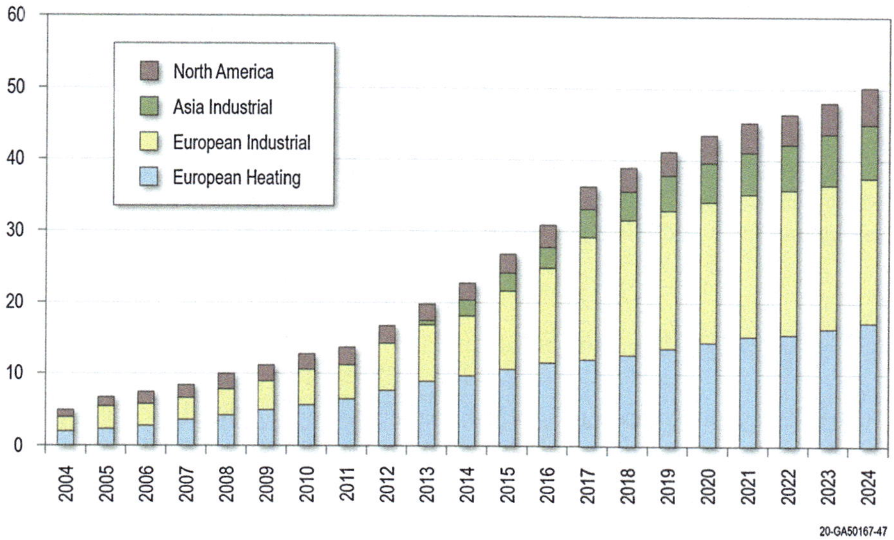

Fig. 6.2 Global pellet demand in millions of tons from 2004 to 2024 (derived from Gemco Energy 2020)

6.2 Global Wood Pellet Consumption

Global wood pellet consumption in 2000 was 1.7 million tons, while in 2015, that number grew to 28 million tons—15 million tons for heat supply and another 13 million tons for use in power plants. As stated previously, most of the pellets produced in North America are consumed in Europe, with about 80% of the overall use in the world (Gemco Energy 2020). The European countries consuming most of these produced pellets are England, Italy, Denmark, Germany, Sweden, Belgium, France, and Austria. For example, the Drax power plant in the United Kingdom has converted its boilers to use pellets to create electricity instead of burning coal. One of the primary reasons for doing this is that many European governments have begun to charge high rates for taxes on coal, oil, and gas, but provide rebates on those using renewable energy sources instead (Gemco Energy 2020).

In addition, many countries in Asia, such as Japan, South Korea, and China, are beginning to see the advantages of using biomass pellets. The pellet consumption in these countries is growing rapidly. One of the major driving factors for this is the increasing desire to replace fossil-based coal sources with biomass-based densified biomass. Due to its incredibly favorable political and environmentally friendly nature, using clean-burning pellets in its plants instead of creating pollution from burning coal is much desired. The other environmental factor influencing the demand for pellets in Asia has been global warming, which has been responsible for higher winter temperatures throughout the continent, thereby driving the price of crude oil down significantly.

In North America, the desire to use pellets for heating has become strongly correlated to the price of heating oil and propane. The demand for wood pellets and pellet stoves has gone up along with the price of oil for heating. Government policy also has an impact on the industrial wood pellet market as follows: (a) Canada has raised the price of carbon to restrain carbon emissions; and (b) the Alberta province has enacted laws to stop power generation from coals by 2030. As such, North America now consistently consumes about 11% of the total pellets produced throughout the world (Gemco Energy 2020).

6.3 Pellet Quality and Standards

High pellet quality is necessary to obtain constant and optimal heating. With changes in process conditions, densified products with varied quality are often produced. Densified products, such as pellets and briquettes with different physical properties and chemical compositions, are available. The densified product physical properties and chemical compositions are influenced based on the type of feedstock, growing conditions, densification process conditions, and binders used to produce high-quality densified products. With such a difference in the quality of the densified products that are available throughout the market, there is a need to have standards that can help the power producers or individual pellet stove owners that they are using the right quality of fuel in their equipment. Using poor quality densified products can have an impact on power production efficiency. In addition, disturbing the combustion process can damage appliances and increase maintenance costs. Pellet quality is much more important for smaller appliances like boilers or stoves because when using pellets at a smaller scale, these systems have much less tolerance. With larger systems, however, pellet quality in terms of physical properties may not cause problems, but chemical compositions such as ash (alkali and alkaline earth metals), chlorine, sulfur, etc., can damage boilers and other biopower generating systems. Tumuluru et al. (2012) discussed how various physical properties and chemical compositions could have a significant impact on biopower generating systems, as shown in Table 6.1.

In general, biomass ash composition is highly variable. Normally, herbaceous biomass, such as straws, residues, and dedicated energy crops used for biopower generation, has a high amount of ash, whereas woody biomass has a lower ash content. For biopower generation, woody biomass is the most preferred feedstock for co-firing and biopower generation. Table 6.2 shows the ash content of various woody and herbaceous biomass sources. According to Tumuluru et al. (2012), the ash content and composition of biomass typically depends on plant growth conditions, such as soil type, water quality, and the harvesting methods followed. Tumuluru et al. (2016) discussed how harvesting methods would impact the ash content in corn stover used for biofuels production.

Table 6.1 Impact of physical and chemical characteristics of biomass and their effects on co-firing

Physical properties	**Effects**
Moisture content	High and variable moisture results in dry-matter losses, spontaneous combustion, and reduce calorific value.
Volatile content	Higher volatile content influences the local fuel-air mixing and combustion considerations, soot formation, and radiation phenomenon impacts.
Bulk density	Bulk density influences the logistics (e.g., storage, transport, handling costs).
Ash content	Higher ash content results in particle emissions, ash utilization and disposal, heat transfer surface fouling, combustion chemistry interactions, SCR catalyst dusting, and poisoning.
Particle size/shape distributions	Particle size, shape, and distribution influence (a) fuel feeding system, (b) combustion chemistry, (c) drying properties, (d)dust formation, and (e) safety (explosion considerations).
Chemical compositions	
Carbon and hydrogen	Positive effect on gross calorific value.
Oxygen	Negative effect on gross CV and peak flame temperature.
Chlorine	Result in corrosion, enhanced mercury capture (conversion of elemental mercury to particulate chlorides).
Nitrogen	Result in NOx, N_2O, and HCN emissions (potentially reduces fuel/prompt).
Sulfur	Result in SOx emissions and corrosion generally reduced.
Fluorine	Hydrogen fluoride emissions and corrosion.
Potassium and sodium	Corrosion, lower ash melting temperature, aerosol formation, and catalyzes coal char reactions.
Magnesium	Hotter ash melting temperature.
Calcium	Ash utilization and catalyzes coal char reactions.
Phosphorus	Hotter ash melting temperature.
Heavy metals	Emissions of pollutants, aerosol formation, ash utilization, and disposal issues.

Table 6.2 Ash content of different biomass fuels and coal (derived from Tumuluru et al. 2012)

Biomass fuel	Ash content (% wt, w.b.)
Bark	5–8
Wood chips with bark (forest)	1–2.5
Wood chips without bark (industrial)	0.8–1.4
Sawdust	0.5–1.1
Waste wood	3–12
Starch and cereals	4–12
Miscanthus	2–8
Coal	5–45 (8.5–10.5 on average)

Ash content in biomass, in general, comes from one of four different forms: (a) easily leachable salts; (b) inorganic elements; (c) minerals; and (d) inorganic materials. Biomass ash content is rich in alkaline metals, which results in the fouling of heat transfer surfaces and ash melting temperatures. For example, chlorine—which is a minor constituent in biomass—results in harmful alkaline and chlorine compounds, thereby causing the fouling of boiler surfaces. Even in small quantities, elemental nitrogen content in biomass can have a significant impact on emissions, thus resulting in the formation of thermal nitrogen oxides (NOx). NOx contains hydrocarbons, which photochemically leads to the formation of ozone—a lung and eye irritant that has become a significant issue in urban environments (Robinson et al. 1998).

Another critical characteristic of biomass is its moisture content, which impacts thermal efficiency. If the biomass has a higher amount of moisture content, it takes a much higher amount of heat to evaporate that moisture. In addition, higher moisture content impacts storage stability, which leads to microbial and fungal degradation and results in dry matter losses that decrease the calorific value (CV) of densified biomass. In general, a higher CV is desirable for power generation. Typically, for densified woody biomass such as pellets, CV is in the range of 18.5–19.5 MJ/kg, whereas for herbaceous biomass, CV is in the range of 17.5–18.5 MJ/kg. Other physical properties, such as bulk density and percent fines, also impact transportation efficiencies and revenue to pellet producers. Therefore, getting good-quality pellets is very important for different applications. Table 6.3 provides a list of some of the quality standards that can have a significant impact on biopower generation.

6.4 International Standards

There are International, European, and American standards for solid fuels. European standards are defined by the European Committee for Standardization (CEN). International standards were developed by the International Standards Organization (ISO). The Pellet Fuels Institute (PFI) developed standards for pellets in the United States. Table 6.4 provides a brief look at the PFI standard specification. There are three grades of pellets—premium, standard, and utility—according to PFI. These standards do not include information regarding heavy metals. The major difference in these three quality pellets is durability, percent fines, inorganic ash, and moisture content.

6.5 CEN Standards

Table 6.5 shows the individual standards used by the European countries, such as Austria, Sweden, the United Kingdom, France, and Denmark. A special environment label for solid biomass fuels was devised by the Austrian Federal Ministry for the Environment. Natural wood (e.g., sawdust, shavings, etc.) is allowed for making

Table 6.3 Impact of various biomass densification specifications on biofuels power generation

Densified product specification	Impact
Ash content	The residue left after the complete combustion of biomass is ash. The pellet will burn more efficiently if the ash content is low. Higher ash content can lead to more cleaning and maintenance.
Ash melting temperature	The ash melting temperature influences the slag formation. Higher slag formation influences the combustion efficiency, and blocking of the equipment can lead to higher maintenance costs.
Length, diameter, and bulk density	Pellet dimensions such as length (mm), diameter (mm), and bulk density (kg/m^3) are important parameters. They impact energy density, ease of feeding, and can help to increase transportation efficiency. Higher energy density increases boiler efficiency.
Fines	Lower fines in the densified biomass are desirable as it impacts the storage, handling, transportation, emissions during combustion, and maintenance. Higher fines in the pelleted biomass can result in revenue loss to the pellet producers as they are paid for the whole pellet. In addition, fines can result in safety issues, such as fire hazards and more pellet off-gassing.
Heating value	Heating value is the amount of heat in the biomass in a certain mass. It is generally expressed as MJ/kg or kWh/kg, which identifies it as having a higher heating value or a lower heating value. Lower heating value is more relevant as it does not consider moisture in biomass.
Moisture content	Moisture content reflects the quantity of water in the pellets. The lower moisture in wood pellets results in a higher heat output per kg., in general, pellets have about 5–7% (w.b.) moisture (Tumuluru et al. 2010a).
Mechanical durability	Durability measures pellet integrity. It refers to the resistance of the pellets to various unit operations it is exposed to during storage, handling, and transportation. In general, higher durability is preferred as they are less prone to breakdown and crumbling during storage, transport, and handling in fuel feed augers, and produces fewer fines.
Nitrogen, sulfur, and chlorine	These parameters impact the polluting emissions and corrosion of boiler surfaces.
Heavy metals	Impact polluting emission, combustion behavior, and ash content.

solid fuels. The use of packaging, coatings, glues, chipboard, or fiberboard residues is not allowed in solid fuel production. Swedish wood pellet trade body Pellsam set up by manufacturers of pellet heating equipment developed the standards for solid biofuels. The United Kingdom pellet market introduced wood pellets to the country with its Department of Trade and Industry (DTI). British BioGen developed Codes of Good Practice for biofuel pellets and pellet-burning room heaters <15 kW. These were developed as interim standards that will be further superseded with European Standards. The French International Association for Bioenergy Professionals (ITEBE) created a Pellet Club with the aim of promoting high-quality solid fuels. The aim of this body is not only to promote the standards but to provide specific advice to determine the quality of pellets for various applications as well. In

Table 6.4 Standard specification for residential/commercial densified fuel based on Pellet Fuel Institute (PFI) (derived from PFI 2008)

Fuel property	PFI premium	PFI standard	PFI utility
Normative information—mandatory			
Bulk density (lb/ft^3)	40.0–46.0	38.0–46.0	38.0–46.0
Diameter (in.)	0.230–0.285	0.230–0.285	0.230–0.285
Diameter (mm)	5.84–7.25	5.84–7.25	5.84–7.25
Pellet durability index (PDI)	≥96.5	≥95.0	≥95.0
Percent fines (at the mill gate)	≤0.50	≤1.0	≤1.0
Inorganic ash (%)	≤1.0	≤2.0	≤6.0
Length (% greater than 1.50 in.)	≤1.0	≤1.0	≤1.0
Moisture (%)	≤8.0	≤10.0	≤10.0
Chloride (ppm)	≤300	≤300	≤300
Informative only—not mandatory			
Ash fusion	NA	NA	NA
Heating value	NA	NA	NA

Denmark, general environment labels such as Svanenmärket (the Scandinavian Blue Angel) or Die Blumme are used for pellets. Table 6.5 shows the comparison of various pellet standards used in Europe by different countries prior to CEN. All individual standards developed by various countries in Europe have since been replaced with the one developed by CEN (CEN 2014).

6.6 European Common Standard

The European Common Standard (CEN) solid biofuels standard was established to avoid the ambiguity that had arisen due to the various standards previously available. Solid biofuels technical specifications and testing methods are developed by CEN/TC 335. Here are some of the other standards developed by CEN for the classification and properties of solid biofuels:

1. The classification of biomass-based origin from CEN/TS 14961, which is broken down into the following subcategories: wood biomass, herbaceous biomass, fruit biomass, and blends and mixtures (see Table 6.6).
2. CEN technical specifications for pellets and briquettes are given in Tables 6.7 and 6.8.

Table 6.5 National standards and codes of good practice for fuel pellets in various European Countries (derived from Tumuluru et al. 2010a)

Specification	Austria ÖNORM M7135		Sweden SS 18 71 20			Germany DIN 51731/DIN plus	Italy CTI – R 04/5				British BioGen/UK codes of good practice	
	Holzpresslinge	Rindenpresslinge	Group 1	Group 2	Group 3	5 Längenklassen sonst gleich [cm]	4 categories according origin and with or without pressing aids A, no additives A, with additives B C					
	Pellets	Briketts:										
Size	Maximum diameter 4–20 mm and 100 mm length	Maximum 20–120 mm diameter and 400 mm length	Maximum 4 mm dia	Maximum 5 mm dia	Maximum 6 mm dia	Length diameter HP1 >30 > 10 HP2 15–30 6–10 HP3 10–15 3–7 HP4 <10 1–4 HP5 <5 0.4–1	6 & 8 mm	6 & 8 mm	6 & 8 mm	10–25 mm	<4 mm–20 mm	>10 mm–<20 mm
Bulk density			≥600 kg/m^3**	≥500 kg/m^3	≥500 kg/m^3		620–720 kg/m^3	620–720 kg/m^3	620–720 kg/m^3	≥550 kg/m^3	>600 kg/m^3**	>500 kg/m^3**
Fines in% <3 mm			≤0.8	≤1.5	≤1.5						<0.5%	<0.5%
Unit density	≥1.0 kg/dm^3	≥1.0 kg/dm^3				1–1.4 g/cm^3						>40 pounds/ cubic ft. +
Moisture content	≤12%	≤18%	≤10%	≤10%	≤12%	<12%	≤10%	≤10%	≤10%	≤15%	≤10%	≤10%
Ash content	≤0.5%*	≤6.0%*)	≤0.7%	≤1.5%	>1.5%	>1.5%	≤0.7%	≤0.7%	≤1.5%	To be stated	< 1%, <3% or 6	<1%, <3% or 6
Caloric value	≥18.0 MJ/kg*	≥18.0 MJ/kg*)	≥16.9 MJ/kg ≥4.7 kWh/kg	≥16.9 MJ/kg ≥4,7 kWh/kg	≥16.9 MJ/kg ≥4.7 kWh/kg	17.5–19.5 MJ/kg ***	≥16.9 MJ/kg	≥16.9 MJ/kg	≥16.2 MJ/kg	To be stated	>4.7 kWh/kg	>4.2 kWh/kg
Sulphur	≤0.04%*	≤0.08%*	≤0.08%	≤0.08%	Anges	<0.08	≤0.05%	≤0.05%	≤0.05%	To be stated	<300 ppm	<300 ppm
Nitrogen	≤0.3%*	≤0.6%*				<0.3	≤0.3%	≤0.3%	≤0.3%	To be stated		
Chlorine	≤0.02%*	≤0.04%*	≤0.03%	≤0.03%	Anges	<0.03	≤0.03%	≤0.03%	To be stated		<800 ppm	<800 ppm
Arsenic						<0.8 mg/kg						
Cadmium						<0.5 mg/kg						
Chromium						<8 mg/kg						
Copper						<5 mg/kg						
Mercury						<0.05 mg/kg						
Lead						<10 mg/kg						
Zinc						<100 mg/kg						

(continued)

Table 6.5 (continued)

Specification	Austria ÖNORM M7135		Sweden SS 18 71 20			Germany DIN 51731/DIN plus	Italy CTI – R 04/5				British BioGen/UK codes of good practice	
EOX, extractabl. org,halogens						<3 mg/kg	≤1.0%	≤1.0%	≤1.0%	To be stated		
Fines. Before delivery to customer	Maximum 1%					Maximum 1%						
Additives	Maximum 2% onlynatural		To be stated					None	To be stated			
Ash melting point			Temperature to be stated									
Durability								≥97.7%	≥97.7%	≥95.0%	≥90.0%	

*: Dry basis; **: At factory; ***: Without ash and water

Table 6.6 CEN-based classification of origin and sources of woody biomass (derived from Hahn 2004 and Alakangas 2011)

1.1 Forest and plantation wood	
1.1.1 Whole trees	1.1.1.1 Deciduous wood
	1.1.1.2 Coniferous wood
	1.1.1.3 Short-rotation coppice
	1.1.1.4 Bushes
	1.1.1.5 Blends and mixtures
1.1.2 Stem wood	1.1.2.1 Deciduous
	1.1.2.2 Coniferous
	1.1.2.3 Blends and mixtures
1.1.3 Logging residues	1.1.3.1 Fresh/green (including leaves/ needles)
	1.1.3.2 Dry
	1.1.3.1 Blends and mixtures
1.1.4 Stumps	1.1.4.1 Deciduous wood
	1.1.4.2 Coniferous wood
	1.1.4.3 Short-rotation coppice
	1.1.4.4 Bushes
	1.1.4.5 Blends and mixtures
1.1.5 Bark (from forestry operations)*	
1.1.6 Landscape management woody biomass	
1.2 Wood processing industry, by-products, and residues	
1.2.1 Chemically untreated wood residues	1.2.1.1 Wood without bark
	1.2.1.2 Wood with bark*
	1.2.1.3 Bark (from industry operations)*
	1.2.1.4 Blends and mixtures
1.2.2 Chemically treated wood residues	1.2.2.1 Wood without bark
	1.2.2.2 wood with bark*
	1.2.2.3 Bark (from industry operations)*
	1.2.2.4 Blends and mixtures
1.2.3 Fibrous waste from the pulp and paper industry	1.2.3.1 Chemically untreated fibrous waste
	1.2.3.2 Chemically treated fibrous waste
1.3 Used wood	
1.3.1 Chemically untreated wood	1.3.1.1 Wood without bark
	1.3.1.2 Bark*
	1.3.1.3 Blends and mixtures
1.3.2 Chemically treated wood	1.3.2.1 Wood without bark
	1.3.2.2 Bark*
	1.3.2.3 Blends and mixtures
1.4 Blends and mixtures	

Table 6.7 Technical specifications for pellets, according to CEN (derived from Alakangas 2011)

Origin: according to Table 6.1 in CEN/TS 14961:2004	(1) Herbaceous biomass; (2) woody biomass; (3) fruit biomass; and (4) blends and mixtures
Traded form (see Table 6.2)	Pellets
Normative specifications	
Dimensions (mm) Diameter and length (L)[a]	D06 $\leq$ 6 mm $\pm$ 0,5 mm and L $\leq$ 5 $\times$ diameter; D08 $\leq$ 8 mm $\pm$ 0,5 mm and L $\leq$ 4 $\times$ diameter; D10 $\leq$ 10 mm $\pm$ 0,5 mm and L $\leq$ 4 $\times$ diameter; D12 $\leq$ 12 mm $\pm$ 1 mm and L $\leq$ 4 $\times$ diameter; D25 $\leq$ 25 mm $\pm$ 1 mm and L $\leq$ 4 $\times$ diameter
Moisture (wt. % as received)	M10 $\leq$ 10%; M15 $\leq$ 15%; M20 $\leq$ 20%
Ash (wt. % of dry basis)	A0.7 $\leq$ 0.7 to A6.0 $\leq$ 6.0%; A6.0+ > 6.0% (actual value to be stated)
Sulfur (wt. % of dry basis)[b]	S0.05 $\leq$ 0.05% to S0.10 $\leq$ 0.10%; S0.20+ > 0.20% (actual value to be stated)
Mechanical durability (wt. % of pellets after testing)	DU97.5 $\geq$ 97.5; DU95.0 $\geq$ 95.0; DU90.0 $\geq$ 90.0
Amount of fines (wt. % <3.15 mm)[c]	F1.0 $\leq$ 1.0% to F2.0 $\leq$ 2.0%; F2.0+ > 2.0% (actual value to be stated)
Additives (weight percent of the pressed mass)	Type and content of pressing aids, slagging inhibitors, or any other additives used need to be stated.
Nitrogen (wt. % of dry basis)	N0.3 $\leq$ 0.3% to N3.0 $\leq$ 3.0% N3.0+ > 3.0% (actual value to be stated).
Informative specifications	
Energy density (kWh/m^3 loose) or net caloric value (MJ/kg as received)	The recommendation is to be informed by the retailer.
Bulk density as received (kg/m^3 loose)	The recommendation is to be stated if traded on a volume basis.
Chlorine, Cl (weight of dry basis wt. %)	Recommendation is to be stated in category Cl 0.03, to Cl 0.10, und Cl 0.10+ (actual value to be stated if Cl >0.10%).

[a]Maximum of 20 wt. % of the pellets may have a length of 7. 5 $\times$ diameter
[b]If chemically treated biomass or sulfur-containing additives are used, it is normative to mention sulfur percent
[c]Measured at the last possible place in the pellet production or after production at the factory gate

6.7 ISO Standards

ISO, which is the International Organization for Standardization, is a worldwide union of national standards bodies where the work on the standards is carried out through ISO technical committees. Each interested member of the body has the right to be represented on the committee for which the technical committee has been established. Various international organizations, governmental, and nongovernmental organizations in liaison with the ISO will also take part in the work. ISO-17225-1, ISO-17225-2, and ISO-17225-3 (2014a, 2014b, 2014c) define the solid biofuels/fuels specification and classes. The objective of these standards is

Table 6.8 CEN-based technical specifications of properties for briquettes (derived from the Swedish Standards Institute SIS-CEN/TS 14961:2005)

Origin: according to Table 6.1 in CEN/TS 14961:2004	(1) Herbaceous biomass, (2) woody biomass, (3) fruit biomass, (4) blends and mixtures
Traded form (see Table 6.2)	Pellets
Normative specifications	
Dimensions (mm) used are: Diameter (D) or equivalent (diagonal or crosscut), mm	D40
	D50
	D60
	D80
	D100
	D125
	D125+
Length (L)	L50
	L100
	L200
	L300
	L400
	L400+
Moisture (wt. % as received)	M10
	M15
	M20
Ash (wt. % of dry basis)	A0.7
	A1.5
	A3.0
	A6.0
	A10.0
Sulfur (wt. % of dry basis)[a]	S0.05
	S0.08
	S0.10
	S0.20
	S0.20+
Particle density (kg/dm^3)	DE0.8
	DE1.0
	DE1.1
	DE1.2
Additives (weight percent of the pressed mass)	Type and content of pressing aids, slagging inhibitors, or any other additives used need to be stated.
Nitrogen, N (wt. % dry basis)	N0.3
	N0.5
	N1.0
	N3.0
	M3.0+

(continued)

Table 6.8 (continued)

Informative specifications	
Energy density (kWh/m^3 loose) or net caloric value (MJ/kg as received)	The recommendation is to be informed by the retailer.
Bulk density as received (kg/m^3 loose)	The recommendation is to be stated if traded on a volume basis.
Chlorine, Cl (weight of dry basis wt. %)	Recommendation is to be stated in category Cl 0.03 to Cl 0.10, and Cl 0.10+ (Actual value to be stated if Cl >0.10%).

[a]Maximum of 20 wt. % of the pellets may have a length of 7. 5 × diameter
[b]If chemically treated biomass or sulfur-containing additives are used, then it is normative to mention sulfur percent
[c]Measured at the last possible place in the pellet production or after production at the factory gate

to serve as a tool to enable efficient trading of biofuels by providing clear and unambiguous classification principles for solid biofuels, which can help a good understanding between the buyer and seller. This also acts as a communication tool for equipment manufacturers. According to the ISO standards, solid biomass covers organic and nonfossil material made up of a biological origin that can be used for heat or power generation. "Graded" in the product standards means that the solid biofuels are used in industrial or public sector buildings, small commercial, or household applications. The solid biofuels are traded in different shapes and sizes. The size and shape of the traded solid fuel influences the fuel properties, such as density, which in turn influences the fuel handling and combustion behavior. Table 6.9 indicates the major traded forms and raw materials of solid fuels. Biomass traded forms based on CEN and ISO are the same. The fuel specifications of compressed solid fuels based on ISO, such as pellets and briquettes, are given in Tables 6.10 and 6.11. The main origin-based solid biofuel groups are: (a) woody biomass; (b) herbaceous biomass; (c) fruit biomass; (d) aquatic biomass; and (e) blends and mixtures.

6.7.1 Comparison of CEN, ISO, and PFI Standards

Comparison of biomass pellets and briquettes standards developed by ISO, CEN, and PFI is given in Tables 6.12 and 6.13. In the case of CEN, the moisture is in the range of 10–20% (w.b), whereas for ISO and PFI, the maximum and minimum moisture content values are less than or equal to 5 and 8% (w.b.), respectively. In the case of ash content, the minimum value is less than 1%, whereas the maximum value is ≤6% for PFI standards. The maximum ash content limit is given in the ISO standards (i.e., >10%), whereas the lowest ash limit is given in the CEN standard (<0.7%). In the case of durability, there are only two levels at ≥95.0% and ≥ 96.5% for the PFI standard, respectively, where for the ISO and CEN standards, the highest durability value is ≥97.5%. The lowest value for CEN is ≥90.0%, whereas, for ISO,

Table 6.9 Major traded forms and raw material of solid biofuels (derived from ISO 2014a)

Fuel name	Typical particle size	Common preparation method
Whole tree	>500 mm	No preparation or delimbed
Wood chips	5–100 mm	Cutting with sharp tools
Hog fuel	Varying	Crushing with blunt tools
Stemwood/ roundwood	>100 cm	Cutting with sharp tools
Firewood	5–10 cm	Cutting with sharp tools
Slabs and offcuts	Varying	Cutting with sharp tools
Bark	Varying	Debarking residue from trees (shredded or unshredded)
Bundle	Varying	Lengthwise oriented and bound
Fuel powder	<1 mm	Milling
Sawdust	1–5 mm	Cutting with sharp tools
Shavings	1–30 mm	Planing with sharp tools
Briquettes	Diameter > 25 mm	Mechanical compression
Pellets	Diameter ≤ 25 mm	Mechanical compression
Bales		
(a) Small square bales	(a) 0.1 m^3	(a) Compressed and bound to squares
(b) Big square bales	(b) 3.7 m^3	(b) Compressed and bound to squares
(c) Round bales	(c) 2.1 m^3	(c) Compressed and bound to cylinders
Chopped straw and Energy grass	10–200 mm	Chopped during harvesting or before combustion
Grain or seed	Varying	No preparation or drying except for process operations necessary for storage for cereal grain
Fruit stones or kernel	5–150 mm	No preparation or pressing and extraction by chemicals
Fiber cake	Varying	Prepared from fibrous waste by dewatering
Charcoal	Varying	Charcoal is prepared by the destructive distillation and pyrolysis of biomass
Thermally treated biomass	Varying	Mild pretreatment of biomass at a temperature between 200 and 300 °C for a short time period (e.g., 60 min)

the lowest value is <95.0%, anything lower than that the minimum value needs to be stated. According to the PFI standard, bulk density should be in the range of 608–736 kg/m^3, whereas for CEN, it needs to be included if the pellets are transported based on a volume basis. In the case of the ISO standard, the lowest value is ≥550 kg/m^3, and the maximum is >800 kg/m^3. In the case of nitrogen and chlorine, PFI does not specify, whereas CEN and ISO have maximum and minimum values of ≤0.3% and ≤ 0.2%, and maximum values are >3% and > 0.30%, respectively. Fixed carbon (F.C.), volatiles content, CV, and particle size distribution are not part of CEN and PFI standards, but according to ISO standards, they are included.

Table 6.10 Specifications of properties for pellets (derived from ISO 2014b)

	Traded form	Pellets
Normative	Origin	1: Herbaceous biomass; 2: woody biomass; 3: fruit biomass; 4: aquatic biomass,; 5: blends and mixtures.
	Dimensions	Diameter and length
	D06	6 mm ± 1.0 mm and 3.15 mm < L ≤ 40 mm
	D08	8 mm ± 1.0 mm and 3.15 mm < L ≤ 40 mm
	D10	10 mm ± 1.0 mm and 3.15 mm < L ≤ 40 mm
	D12	12 mm ± 1.0 mm and 3.15 mm < L ≤ 50 mm
	D25	25 mm ± 1.0 mm, and 10 mm < L ≤ 50 mm
	Moisture, M (wt. % as received)	
	M05; M08; M10; M12; M15	≤ 5%; ≤ 8%; ≤ 10%; ≤ 12%; ≤ 15%
	Ash, A (wt. % of dry basis)	
	A0.5 to A10.0; A10.0+	≤ 0.5% to ≤10.0%; > 10.0%(maximum value to be stated)
	Mechanical durability (D.U.) (wt. % of pellets after testing)	
	DU97.5; DU96.5; DU95.0; DU95.0-	≥ 97.5%; ≥ 96.5%; ≥ 95.0%; < 95.0% (minimum value to be stated)
	Amount of fines, F (wt. %, < 3.15 mm) after production when loaded or packed	
	F1.0 to F6.0; F6.0+	≤ 1.0% to ≤6.0%; > 6.0% (maximum value to be stated)
	(wt. % of pressing mass) b	Pressing aids type and content, slagging inhibitors or any other additives used need to be stated
	Bulk density (BD) (kg/m3 as received)	
	BD550; BD580; BD600; BD625; BD650; BD700; BD750; BD800+	≥ 550 kg/m^3; ≥ 580 kg/m^3; ≥ 600 kg/m^3; ≥ 625 kg/m^3; ≥ 650 kg/m^3; ≥ 700 kg/m^3; ≥ 750 kg/m^3; > 800 kg/m^3 (minimum value to be stated)
	Net CV, Q (MJ/kg or kWh/kg as received)	Minimum value to be stated[c]
Normative/ informative	Nitrogen, N (wt. % of dry basis)	
	N0.2 to N3.0; N3.0+	≤ 0.2% to ≤3.0%; > 3.0%(maximum value to be stated)
	Sulfur, S (wt. % of dry basis)	
	S0.02 to S0.20; S0.20+	≤ 0.02% to ≤0.20%; > 0.20% (maximum value to be stated)
	Chlorine, Cl (wt. % of dry basis)	

(continued)

Table 6.10 (continued)

	Traded form	Pellets
	Cl0.01 to Cl0.30; Cl0.30+	≤ 0.01% to ≤0.30%; > 0.30% (maximum value to be stated)
	Fixed carbon, C d (wt. % of dry basis)	Minimum value to be stated
		Normative only for thermally treated biomass pellets
	Volatile matter (V.M.) (wt. % of dry basis	Minimum value to be stated.
		Normative only for thermally treated biomass pellets
	Particle size distribution of disintegrated pellets (wt. % of dry basis)	Values to be stated of pellets for industrial use
Informative	Ash melting behavior (°C)	Should be stated

[a]Amount of pellets longer than 40 mm (or 50 mm) can be 5 wt.%. The maximum length for classes *D*06, *D*08, and *D*10 shall be ≤45 mm. Pellets are longer than 3.15 mm if they stay on a round hole-sieve of 3.15 mm

[b]Maximum amount of additive is 20 wt.% of pressing mass. Type stated (e.g., starch, corn flour, potato flour, vegetable oil, lignin). If the amount is greater, then raw material for pellet is the blend

[c]Minimum value for torrefied or other thermally treated biomass pellets is usual ≥18 M.J./kg.

[d]Fixed carbon (%) is calculated by the following: 100 – [moisture (wt. %) + ash (wt. %) + volatile matter (wt. %)]. All percentages are on the same moisture basis

[e]Special attention should be paid to the ash melting behavior for some biomass fuels, for example, eucalyptus, poplar, short-rotation coppice, straw, miscanthus, and olive stone. It is recommended that all characteristic temperatures (shrinkage starting temperature (SST), deformation temperature (D.T.), hemisphere temperature (H.T.), and flow temperature (F.T.)) in oxidizing conditions should be stated. Pre-ash temperature other than 550 °C should be stated

PFI has not developed standards for other densified products, such as briquettes. CEN and ISO have developed standards for the briquettes. In the case of CEN, the manufacturer has concluded that the length and diameter of the briquette must be included, whereas the ISO standard does not require that information. Moisture content values, according to CEN, should be ≤10 to ≤20% (w.b.), while according to ISO, they should be ≤10 to ≤15% (w.b.). The minimum value of ash content according to CEN and ISO is ≤0.7 and 0.5, respectively, whereas the maximum is >6 and > 10%, respectively. In the case of sulfur, the minimum values according to CEN and ISO are ≤0.05 and 0.02%, respectively, whereas the maximum is >0.2%. The CEN standard does not consider mechanical durability, while ISO has lower and upper values specified (e.g., < 90.0%, minimum value to be stated) and ≥ 95.0%. The particle density of the briquette is included in both standards. Lower values of nitrogen and chlorine for CEN and ISO were ≤ 0.3 and 0.03%, and maximum values were > 3% for nitrogen and > 0.1% for chlorine. According to ISO, it is >0.3%. F.C., volatile content, particle size distribution, and CV are required to be included if chemically treated biomass is used for making densified products, such as briquettes.

The pellets produced using commercial- and laboratory-scale pelleting systems using different woody and herbaceous biomass feedstocks, different feedstock

Table 6.11 Specifications of properties for briquettes (derived from ISO 2014c)

	Traded form	Pellets
Normative	**Origin**	1: Herbaceous biomass; 2: woody biomass; 3: fruit biomass; 4: aquatic biomass; 5: Blends and mixtures
	Dimensions	Diameter and length
	Moisture, M (wt. % as received)	
	M10; M12; M15	≤ 10%; ≤ 12%; ≤ 15%
	Ash, A (wt. % of dry basis)	
	A0.5 to A10.0; A10.0+	≤ 0.5% to ≤10.0%; > 10.0% (maximum value to be stated)
	Particle density, DE (g/cm^3, as received)	
	DE0.8 to DE1.1; DE1.2+	≥ 0.8 g/cm^3 to ≥1,1 g/cm^3; > 1,2 g/cm^3 (maximum value to be stated)
	Additives[a] (wt. % of the press mass)	Need to state the type and content of pressing aids, slagging inhibitors or any other additives
	Net CV, Q (MJ/kg or kWh/kg as received)	The minimum value to be stated[b]
Normative/informative: Only if traded in bulk	**Mechanical durability (DU)** (wt. % of pellets after testing)	
	DU95.0; DU90.0; DU90.0-	≥ 95,0%; ≥ 90,0%; < 90,0% (minimum value to be stated)
Normative: chemically treated biomass	**Nitrogen, N (wt. % of dry basis)**	
Informative: all fuels that are not chemically treated	N0.2; N0.3; N0.5; N0.7; N1.0; N1.5; N2.0; N3.0; N3.0+	≤ 0.2 to ≤3.0%; > 3,0% (maximum value to be stated)
Normative: chemically treated biomass or if additives that have sulfur are used	**Sulfur, S (wt. % of dry basis)**	
Informative: all fuels that which not chemically treated	S0.02 to S0.10; S0.20; S0.20+	≤ 0.02% to ≤0.10%; ≤ 0.20%; > 0.20% (maximum value to be stated)
Normative: chemically treated biomass	**Chlorine, Cl (wt. % of dry basis)**	
Informative: all fuels that are not chemically treated	Cl0.01 to Cl0.30; Cl0.30+	≤ 0.01% to ≤0.30%; > 0.30% (maximum value to be stated)
Normative/informative	**Fixed carbon, C[c] (wt. % of dry basis)**	
	Minimum value to be stated	Normative only for thermally treated biomass briquettes
Normative/informative		

(continued)

Table 6.11 (continued)

	Traded form	Pellets
	Volatile matter (VM) (wt. % of dry basis)	
	Maximum value to be stated	Normative only for thermally treated biomass briquettes
Informative	Ash melting behavior[d] (°C)	Should be stated

[a]The maximum amount of additive is 20 w-% of pressing mass. Type stated as chemical substance (e.g., starch, corn flour, potato flour, vegetable oil, lignin). If the quantity is higher, then raw material for briquette is the blend
[b]Minimum value for torrefied or other thermally treated biomass briquettes is usually ≥18 MJ/kg
[c]Fixed carbon (%) is calculated by the following: 100 – [moisture (w-%) + ash (w-%) + volatile matter (w-%)]. All percentages are on the same moisture basis
[d]Ash melting behavior for some biomass fuels needs special attention, for example, eucalyptus, poplar, short-rotation coppice, straw, miscanthus, and olive stone. It is recommended that all characteristic temperatures (shrinkage starting temperature (SST), deformation temperature (D.T.), hemisphere temperature (H.T.), and flow temperature (F.T.)) in oxidizing conditions should be stated

moisture content values, and different types of mills are provided in Table 6.14. In his studies on the pelleting of woody (e.g., lodgepole pine, eucalyptus) and herbaceous (e.g., switchgrass, corn stover) biomass, Yancey et al. (2013) indicated that pellets made using woody biomass have more bulk density as compared to herbaceous biomass. Tumuluru (2014, 2016) reported that pellet quality made using lodgepole pine changes with moisture content. Table 6.14 shows the bulk density and durability of pellets made using lodgepole pine, eucalyptus, corn stover, switchgrass, and their blends at different moisture content values and mill types. Pelleting from both woody and herbaceous biomass at lower moisture content values of 10–12% (w.b.) helps to meet the high-quality standards established by CEN, PFI, and ISO. Also, pellets made using blends of lodgepole pine, switchgrass, eucalyptus, and corn stover in equal proportions by weight at a moisture content value of 10–12% (w.b) has helped to meet the high-quality standards established for bulk density and durability. Tumuluru et al. (2019, 2020) and worked on producing woody, herbaceous, and blends of biomass briquettes at various moisture content values and using different hammermill screen sizes. These studies also indicated the properties of the briquettes change with process conditions, such as moisture, hammermill screen size, and blend ratios and meet some of the physical properties, such as size, shape, density, and durability, and ash content, as observed in Table 6.15. Figure 6.3 shows the pellets and briquettes produced using woody, herbaceous, and blends of woody and herbaceous biomass.

Biorefineries are currently operating in a way so that they can produce biofuels using both woody and herbaceous biomass. Still, they face significant challenges in doing so in terms of storage, quality, handling, and transportation. Most of the herbaceous biomass used by biorefineries comes in a less convenient form to transport efficiently and with high and variable moisture content values. In addition,

Table 6.12 Some of the comparisons of CEN, ISO, and PFI standards for densified biomass such as pellets briquettes

Densified product type: pellets	CEN	ISO	PFI
Moisture content (wt. % as received)	≤ 10% to ≤20%	≤ 5% to ≤5	≤8.0 to ≤10.0
Ash (wt. % of dry basis)	≤ 0.7% > 6.0%	≤ 0.5% to >10.0% (maximum value to be stated)	≤1.0 to ≤6.0
Sulphur (wt. % of dry basis)	≤ 0.05% > 0.20%	≤ 0.02% to >0.20%	
Mechanical durability (wt. % of pellets after testing)	≥ 97.5 to ≥90.0	≥ 97.5% to <95.0% (minimum value to be stated)	≥95.0 and ≥ 96.5
Amount of fines (wt. % <3.15 mm)	≤ 1.0% to >2.0%	≤ 1.0% to >6.0% (after production when loaded or packed)	≤0.50 to ≤1.0 (at the mill gate)
Bulk density	Need to specify if transported in volume basis	≥ 550 kg/m^3 to > 800 kg/m^3.	608 kg/m^3–736 kg/m^3
Nitrogen, N (wt. % of dry basis)	≤ 0.3% to >3.0%	≤ 0.2% to >3.0%.	
Chlorine, Cl (weight of dry basis wt. %)	0.03%- > 0.10%	≤ 0.01% to >0.30%.	≤0.03%
Fixed carbon, C ^{d} (wt. % of dry basis)	NA	Minimum value to be stated Normative only for thermally treated biomass pellets	N.A.
Volatile matter (VM) (wt. % of dry basis)	NA	Minimum value to be stated Normative only for thermally treated biomass pellets	N.A.
Particle size distribution of disintegrated pellets(w-% of dry basis)	NA	Values to be stated of pellets for industrial use	NA
CV	The recommendation is to be stated at the retail level	Minimum value to be stated	NA

the fibrous nature of biomass makes it difficult to feed into the reactors due to the bridging of particles resulting in conveyor and drop chute jamming, as well as the uneven discharge of the materials into the bioreactors. Tumuluru et al. (2010b) indicated that the densification of biomass into either a pellet or a briquette is critical to improving the storage, handling, and conveying efficiencies of biorefineries.

Changing the moisture content values, such as aiming for higher moisture content, produces pellets with different density and durability values. Typically, high-quality pellets in terms of density and durability are needed for long-distance

Table 6.13 Comparison of CEN and ISO standards for briquettes

Densified product type: briquettes		
	CEN	**ISO**
Length (mm)	≤50 to ≥400 (actual value to be stated)	N/A
Diameter (mm)	25 ≤ D ≤ 40 to ≥125 (actual value to be stated)	N/A
Moisture content (wt. % as received)	≤ 10% to ≤20%	≤ 10%; ≤ 12%; ≤ 15%
Ash (wt. % of dry basis)	≤ 0.7% > 6.0%	≤ 0.5% to >10.0%
Particle density (g/cm^3)	0.8 g/cm^3 to ≥1.20 g/cm^3	≥ 0.8 g/cm3 to >1.20 g/cm^3 (maximum value to be stated).
Sulphur (wt. % of dry basis)	≤ 0.05% > 0.20%	≤ 0.02% to >0.20%
Mechanical durability(wt. % of pellets after testing)	NA	≥ 95,0%
		≥ 90,0%
		< 90,0%
		(minimum value to be stated).
		Note: informative: only if the briquettes are traded in bulk
Amount of fines (wt. % <3.15 mm)	N/A	N/A
Bulk density	The recommendation is to be stated if traded by volume basis	N/A
Nitrogen (N) (wt. % of dry basis)	≤ 0.3%; > 3.0% (actual value to be stated)	≤ 0.2% to >3.0% (maximum value to be stated)
		Normative: chemically treated biomass
		Informative: fuels which are not chemically treated
Chlorine (Cl) (weight of dry basis wt. %)	0.03%; >0.10%(actual value to be stated)	≤ 0.01% to >0.30%
		Normative: chemically treated biomass
		Informative: fuels that are not chemically treated
Fixed carbon (FC) (wt. % of dry basis)	N/A	Minimum value to be stated
		Normative: only for thermally treated biomass pellets

(continued)

Table 6.13 (continued)

Densified product type: briquettes		
	CEN	**ISO**
Volatile matter (VM) (wt. % of dry basis)	N/A	Minimum value to be stated **Normative**: only for thermally treated biomass pellets
Particle size distribution of disintegrated pellets (wt. % of dry basis)	N/A	Values to be stated of pellets for industrial use
CV	The recommendation is to be stated at the retail level.	Minimum value to be stated

transportation, such as shipping the material to Europe and Asia. Transporting pellets within the United States may not demand a very high-quality pellet in terms of density and durability, as the quality is directly related to the cost. For example, if the pellets are transported by truck, which is limited by weight, the maximum density desirable to fill the truck full is about 400 kg/m^3. A process that can produce pellets with different density and durability values can help reduce pelleting costs, as well as help to meet different transportation scenarios. Studies conducted by Lamers et al. (2015) indicated that producing pellets at higher moisture content values reduces pellet production costs by about 50%, as compared to the conventional method followed by the industry. Currently, costs associated with the conventional pelleting process (e.g., size reduction, drying, densification) does not off-set transportation costs. There are other advantages in using pelleted feedstocks, such as handling and storage benefits, which have been quantified in terms of cost-savings. Reducing pelleting and transportation costs, as well as quantifying the storage, quality, and handling benefits, can make pelleting a cost-effective solution for biorefineries.

6.8 Conclusions

Europe is a significant pellet producer and supplier, taking up to 51% of the market share in 2016. In North America, the United States produces about 27% of the world's total wood pellet production, whereas Canada produces about 7%. The province of British Columbia, Canada, is a major pellet producer. Many factors influence pellet production demand, such as government policy (i.e., high government levy taxes on gas, oil, and costs), environmental factors, such as winter temperatures, and the overall cost of heating oil. Various physical properties, such as density and moisture impact handling, storage, transportation, and conversion efficiencies and chemical compositions, impact biopower generation and biofuels production. Densification of biomass into pellets and briquettes helps to overcome these limitations. The densified products that are produced using various woody and

Table 6.14 Comparison of bulk density and durability of pellets made using woody, herbaceous, and blends with the existing international standards

Physical property	Moisture content value (%, w.b.)	Durability (%)	Diameter (mm)	Bulk density (kg/m^3)
CEN standard	M10 ≤ 10%	DU97.5 ≥ 97.5 DU95.0 ≥ 95.0 DU90.0 ≥ 90.0	D06 ≤ 6 mm ± 0.5 mm D08 ≤ 8 mm ± 0.5 mm	It is recommended to state the bulk density if traded in volume basis
Pellet fuel institute (utility standard)	≤10	>95	≤ 10	576–736
ISO standards	5–15	≥ 97.5; ≥ 96.5; ≥ 95.0 and < 95.0 (minimum value to be stated).	6–25	550–800
Commercial wood pellets quality (Tumuluru et al. 2010a)	3.44–6.31	>97.5%	6 mm ± 0.5 mm	787–880
Corn stover pellets made at 10–12% moisture content (Yancey et al. 2013) (pellet mill used: 5 ton/h capacity)	<8	98.5	6 mm ± 0.5 mm	645
Switchgrass pellets made at 10–12% moisture content (Yancey et al. 2013) (pellet mill used: 5 ton/h capacity)	<8	97.5	6 mm ± 0.5 mm	716
Lodgepole pellets made at 10–12% moisture content (Yancey et al. 2013) (pellet mill used: 5 ton/h capacity)	<8	96.8	6 mm ± 0.5 mm	712
Eucalyptus pellets made at 10–12% moisture content (Yancey et al. 2013) (pellet mill used: 5 ton/h capacity)	<8	95.4	6 mm ± 0.5 mm	665
Four blend (corn stover: lodgepole pine: switchgrass: eucalyptus) pellets made at 10–12% moisture content (Yancey et al. 2013) (pellet mill used: 5 ton/h)	<8	97.9	6 mm ± 0.5 mm	730
Lodgepole pine wood pellets made at 33–39% (w.b.) moisture content (Tumuluru 2015) (flat die pellet mill: 40 lb./h)	<10	93–96.5	8 ± 0.5 mm	368–562

(continued)

Table 6.14 (continued)

Physical property	Moisture content value (%, w.b.)	Durability (%)	Diameter (mm)	Bulk density (kg/m^3)
Corn stover pellets made at 28–38% (w.b.) moisture content (Tumuluru, 2016) (flat die pellet mill: 40 lb./h)	<10	83–96	8 ± 0.5 mm	399–626
Switchgrass + lodgepole pine blend pellets made at 20–30% (w.b.) moisture content (50:50) ratio (flat die pellet mill: 40 lb./h)	<10	80–95	8 ± 0.5 mm	407–618
Switchgrass + lodgepole pine blend pellets made at 20–30% (w.b.) moisture content (25:75) ratio (flat die pellet mill: 40 lb./h)	<10	76–96	8 ± 0.5 mm	350–552
Switchgrass + lodgepole pine blend pellets made at 20–30% (w.b.) moisture content (75:25) ratio (flat die pellet mill:40 lb./h)	<10	77–91	8 ± 0.5 mm	459–602

Table 6.15 Comparison of briquette physical properties (density and durability) produced using woody, herbaceous, and blends of woody and herbaceous biomass blends according to ISO standards

Physical property	Moisture content value (%, w.b.)	Durability (%)	Diameter (mm)	Length (mm)	Bulk density (kg/m^3)
CEN standard	M10: ≤10% M15: ≤15%	DU97.5 ≥ 97.5 DU95.0 ≥ 95.0 DU90.0 ≥ 90.0	D40: 25 ≤ D ≤ 40 D50 D60 D80 D100 D125 D125+	L50: ≤50 L100 L200 L300 L400 L400+	It is recommended to state the bulk density if traded in volume basis
ISO standards	M10 ≤ 10% M12 ≤ 12% M15 ≤ 15%	DU95.0 ≥ 95.0 DU90.0 ≥ 90.0 DU90.0 < 90.0 (minimum value to be stated)	Diameter, height, width, and length need to be stated		It is recommended to state the bulk density if traded in volume basis
Corn stover briquettes Moisture 12–18% (w.b.); Hammermill screen size:4.8–19.05 mm	<11% (w.b.)	92–99	50.4	50–100	393–489
Switchgrass briquettes Moisture 12–18% (w.b.); Hammermill screen size:4.8–19.05 mm	<11% (w.b.)	79–98	50.4	50–100	331–497
Lodgepole pine briquettes Moisture 12–18% (w.b.); Hammermill screen size:4.8–19.05 mm	<11% (w.b.)	86–97	50.4	50–100	368–489

(continued)

Table 6.15 (continued)

Physical property	Moisture content value (%, w.b.)	Durability (%)	Diameter (mm)	Length (mm)	Bulk density (kg/m^3)
Woody and herbaceous biomass blends					
Switchgrass: corn stover Moisture content: 12–18% (w.b.); Hammermill screen size: 4.8–12.7 mm	<12% (w.b.)	98.5–95	50.4	50–70	356–457
Lodgepole pine: corn stover Moisture content: 12–18% (w.b.); Hammermill screen size: 4.8–12.7 mm	<12% (w.b.)	95–98	50.4	60–70	340–433
Lodgepole pine: switchgrass Moisture content: 12–18% (w.b.); Hammermill screen size: 12.7 mm	<12% (w.b.)	97.5	50.4	66	340
Lodgepole pine: switchgrass: corn stover Moisture: 12–18% (w.b.); Hammermill screen size: 12.7 mm	<12% (w.b.)	95.5–97.6	50.4	61–73	420–439

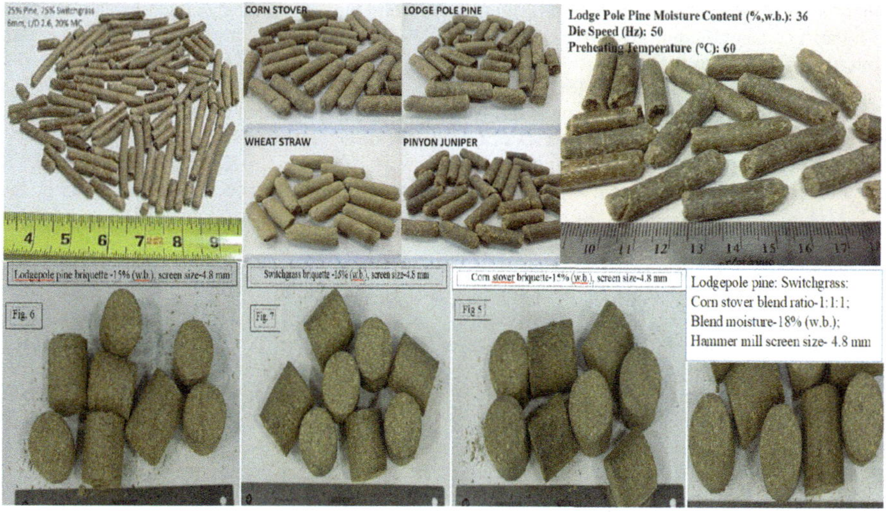

Fig. 6.3 Pellets and briquettes produced using different biomass feedstocks (Tumuluru 2019; Tumuluru, 2018; Tumuluru, 2019a; Tumuluru and Fillerup, 2020)

herbaceous biomass are completed according to the respective international standards. The United States is governed by the PFI standards, while Europe follows the standards developed by CEN and ISO. ISO standards are also used for comparing pellet and briquette quality using woody, herbaceous, and blends of biomass. It can be concluded that the ISO standards are more flexible in terms of accommodating various types of biomass for energy applications.

References

Alakangas E (2011) European standards for fuel specification and classes of solid biofuels. In: Grammelis P (ed) Solid Biofuels for Energy. Green energy and technology. Springer, London

Cocchi M, Nikolaisen L, Junginger M, Goh CS, Heinimo J, Bradley D, Hess R, Jacobson J, Ovard LP, Thran D, Hennig C, Deutmeyer M, Schouwenberg PP, Marchal D (2011) Global wood pellet industry market and trade study. IEA Bioenergy—Task 40: Sustainable International Bioenergy Trade [Online]. Available at: http://task40.ieabioenergy.com/wp-content/uploads/2013/09/t40-global-wood-pellet-market-study_final_R.pdf. Accessed 16 July 2020

EPC Survey (2017) IEA Bioenergy Task 40. FAO, CREIA

European Committee for Standardization (CEN) (2014) Solid biofuels. CEN/TC 335 [Online]. Available at: https://standardsceneu/dyn/www/f?p=204:7:0::::FSP_ORG_ID:19930&cs=17158638AB0C35D5E52A369017E54A1D6. Accessed 16 July 2020

Gemco Energy (2020) Wood Pellet Market in Europe, North America, and Asia. GEMCO Energy Machinery Co, Ltd [Online] Available at: http://wwwgemcopelletmills/wood-pellet-markethtml Accessed 12 June 2020

Hahn B (2004) Existing guidelines and quality assurance for fuel pellets: Pellets for Europe project. UMBERA, Umweltorientierte Betriebsberatungs-, Forschungs- und Entsorgungs-Gesellschaft m.b.H., Schießstattring 25/4, Pölten, Austria

International Standards Organization (ISO) (2014a) Solid biofuels: Fuel specifications and classes: Part 1: General requirements. ISO 17225-1:2014. ISO, Geneva, Switzerland
International Standards Organization (ISO) (2014b) Solid biofuels: Fuel specifications and classes: Part 2: Graded wood pellets. ISO 17225-2:2014. ISO, Geneva, Switzerland
International Standards Organization (ISO) (2014c) Solid biofuels: Fuel specifications and classes: Part 3: Graded wood briquettes. ISO 17225-3:2014. ISO, Geneva, Switzerland
Lamers P, Roni MS, Tumuluru JS, Jacobson JJ, Cafferty KG, Hansen JK, Kenney K, Teymouri F, Bals B (2015) Techno-economic analysis of decentralized biomass processing depots. Bioresour Technol 194:205–213
Pellet Fuel Institute (PFI) (2008) Standard specification for residential/commercial densified fuel. PFI, Seattle. [Online]. Available at: https://www.pelletheat.org/assets/docs/standardspecificationwithcopyright.pdf Accessed 12 June 2020
Robinson AL, Baxter LL, Junker H, Shaddix C, Freeman M, James R, Dayton D (1998) Fireside issues associated with coal-biomass cofiring, NREL/TP-570-25767, December 1998. National Renewable Energy Laboratory, Golden
Swedish Standards Institute (2005) Solid biofuels: fuel specifications and classes, SIS-CEN/TS 14961:2005. Swedish Standards Institute, Stockholm. [Online] Available at: https://wwwsisse/api/document/preview/39467/. Accessed 12 June 2020
Tumuluru JS (2014) Effect of process variables on the density and durability of the pellets made from high moisture corn stover. Biosyst Eng 119:44–57
Tumuluru JS (2015) High moisture corn stover pelleting in a flat die pellet mill fitted with a 6-mm die: physical properties and specific energy consumption. Energy Sci Eng 3:327–341
Tumuluru JS (2016) Specific energy consumption and quality of wood pellets produced using high moisture lodgepole pine grind in a flat die pellet mill. Chem Eng Res Des 110:82–97
Tumuluru JS (2019) Effect of moisture content and hammer mill screen size on the briquetting characteristics of woody and herbaceous biomass. KONA Powder Part J 36:241–251
Tumuluru JS, Fillerup E, Kane J, Murray DJ (2019) Biomass engineering: size reduction, drying, and densification of high moisture biomass. Feedstock Supply and Logistics Platform, US Department of Energy, Project Peer Review, DOE Bioenergy Technologies Office, 6 March, Denver, CO, USA [Online]. Available at: https://www.energy.gov/sites/prod/files/2019/04/f61/Size%20Reduction%2C%20Drying%20and%20Densification%20of%20High%20Moisture%20Biomass_NL0026654.pdf. Accessed 8Apr 2020
Tumuluru JS, Fillerup E, Kane JJ, Murray D (2020) Advanced imaging techniques to understand the impact of process variables on the particle morphology in a corn stover pellet. Chem Eng Res Des 161:130–145
Tumuluru JS, Wright CT, Kenney KL, Hess JR (2010a) A review on biomass densification technologies for energy applications, INL/EXT-10-18420. Idaho National Laboratory, Idaho Falls
Tumuluru JS, Sokhansanj S, Lim CJ, Bi X, Anthony L, Staffan M, Sowlati T, Ehsan O (2010b) Quality of wood pellets produced in British Columbia. Appl Eng Agric 26(6):1013–1020
Tumuluru JS, Hess JR, Boardman RD, Wright CT, Westover TL (2012) Formulation, pretreatment, and densification options to improve biomass specifications for cofiring high percentages with coal. Ind Biotechnol 8(3):113–132
Tumuluru JS, Conner CC, Hoover AN (2016) Method to produce durable pellets at lower energy consumption using high moisture corn stover and a corn starch binder in a flat die pellet mill. J Vis Exp 112:e54092
Will S (2009) An overview of the wood pellet industry: 'A British Columbia perspective.' WOOD 493. A Report Submitted in Partial Fulfillment of the Requirements for the Degree of Bachelor of Science in Wood Products Processing in the Faculty of Forestry at the University of British Columbia [Online]. Available at: https://open.library.ubc.ca/cIRcle/collections/undergraduateresearch/52966/items/1.0103096 Accessed 16 July 2020
Yancey NA, Tumuluru JS, Wright CT (2013) Drying, grinding, and pelletization studies on raw and formulated biomass feedstocks for bioenergy applications. J Biobased Mater Bioenergy 7(5):549–558

Index

A

Advanced grinding concepts, 117–119
Agglomerator, 37, 44, 46, 124
Aquatic biomass, 3, 174, 176, 178
Ash content, 4, 5, 8, 9, 18, 24, 164, 165, 167, 169, 174, 177, 179

B

Belt dryers, 13, 89, 110, 111, 113
Belt filter press, 114
Biochemical composition, 11, 12, 15–18
Biochemical conversion, 9, 11, 46, 124–126, 132, 141, 145, 146, 157
Biomass, 1, 23, 63, 85, 123, 161
  composition, 50, 51, 53, 55, 56, 138
  conversion, 157
  limitations, 15, 17, 18, 124
  types, 2, 6, 89, 150
Biopower, 4, 5, 15, 46, 124, 144, 152, 155, 156, 164, 166, 182
Blends, 2, 3, 6, 53, 88, 89, 127, 128, 168, 171–174, 176–179, 183–187
Briquette press, 15, 18, 24, 32–34, 43, 46, 57, 78, 88, 124, 126
Briquettes, 15, 33, 34, 36, 41, 44–46, 48, 51, 53, 55, 107, 135, 136, 146, 164, 168, 173–175, 177–182, 185, 187
Bulk density, 5, 13–18, 23, 24, 38, 49, 53–55, 57, 79, 80, 89, 90, 101, 103, 117, 124, 126, 127, 135, 146, 165–169, 172, 174–176, 179–181, 183–186

C

Calorific values (CVs), 9–12, 16, 17, 55–57, 124, 136, 141, 146, 150, 165, 166, 175–178, 180, 182
Centrifuge thickening and dewatering system, 114
Chemical composition, 5, 15, 17, 18, 25, 35, 50, 51, 56, 85, 124, 126, 127, 137, 141, 164, 165, 182
Compression process models, 63
Conventional densification process, 87, 89
Cooper-Eaton model, 64, 66, 68, 69, 81

D

Densification, 14, 23, 64, 87, 124, 164
  process variables, 41, 45, 47, 48, 50, 51, 57
  systems, 15, 18, 23–57, 63, 124
Densified biomass physical properties, 51, 53–56
Densified products, 16, 29, 36, 44, 45, 47–51, 55, 57, 63, 101, 105, 123–157, 161–187
Dewatering technologies, 109, 110, 112–114, 116
Durability, 16, 45, 47–49, 51, 55–57, 78, 79, 89, 90, 101, 103, 104, 132, 141, 144, 166, 167, 170, 172, 174, 176–181, 184, 185
Dynamic mechanical analysis (DMA), 26, 27, 76

© Springer Nature Switzerland AG 2020
J. S. Tumuluru, *Biomass Densification*,
https://doi.org/10.1007/978-3-030-62888-8

E
Energy consumption, 1, 13, 40, 42–46, 48, 55, 57, 78, 88, 92, 95, 101, 103, 107, 108, 111, 116, 117, 119, 132, 144
European Committee for Standardization (CEN), 16, 101, 166, 168, 172–174, 177, 179–181, 183, 185, 187
Evolutionary algorithms, 78, 80

F
Feedstock properties, 18, 45, 47, 49, 56, 57, 126, 132
Flat die pellet mill, 29, 31, 42, 49, 51, 54, 56, 75, 183, 184
Forest and plantation wood, 171
Fractional milling, 91–96, 98, 101, 103, 105–107, 119, 125
Fruit biomass, 3, 168, 172–174, 176, 178

G
Gasification, 4, 6, 40, 123, 126, 147–154
Geometric mean particle length, 14, 15
Grinding energy, 5, 15, 18, 23, 92, 94, 95, 98, 101, 105, 117–119, 124, 126
Grind moisture, 98, 103, 104

H
Herbaceous biomass, 3, 6, 8, 9, 11–15, 18, 23, 46, 51, 53, 85, 87–89, 109, 118, 119, 124, 126, 157, 161, 164, 166, 168, 172–174, 176–179, 182, 185, 186
Holm's model, 69, 75

I
International standards, 161–187
International Standards Organization (ISO), 2, 15, 16, 166, 172, 174, 176–183, 185, 187

J
Jones model, 66, 68, 69, 81

K
Kawakita and Lüdde model, 64, 67–71, 73–76, 81

L
Low- and medium-temperature drying technologies, 110

M
Mixtures, 2, 3, 64, 66, 127, 147, 148, 168, 171, 173, 174, 176, 178
Moisture content, 5, 24, 68, 85, 165
Moisture management, 91, 92
Municipal solid waste (MSW), 4, 6, 7, 10, 12, 14, 31, 45, 85, 88, 109, 110, 155

O
Optimization, 63–81

P
Packed moving bed (PMB) dryer, 117
Panelli-Filho model, 68, 69, 72
Particle binding mechanism, 24
Particle handling, 86–119
Pellet, 14, 24, 63, 88, 124
Pellet Fuel Institute (PFI), 15, 17, 101, 166, 168, 174, 177, 179–181, 183, 185, 187
Pellet mills, 15, 18, 24, 29–33, 41, 42, 44, 46, 49, 51–54, 56, 57, 75, 76, 78, 88, 101–103, 108, 124, 126, 132, 156, 183
Pellet quality and standards, 164–166
Pellets, 161
Physical properties, 5, 13–16, 18, 35, 51, 53–57, 85, 87, 88, 117, 119, 124, 132, 136, 137, 141, 164, 166, 179, 182, 185
Plate and frame filter press (PFFP), 114, 115
Preconditioning, 30
Process variables, 23–57, 63, 77, 78, 80, 94, 98, 126, 131, 135
Product properties, 78, 95
Proximate and ultimate composition, 5, 7, 8, 16
Pyrolysis, 4, 123, 126, 147–154, 175

R
Response surface methodology, 77–81, 98
Ring die pellet mill, 29, 31–33, 42, 52, 53, 76, 103
Roller press, 36, 37, 44, 46
Rotary press, 115, 116

S
Scale-up methodology, 69, 71, 72, 74–76
Scanning electron microscope (SEM), 27–29, 57
Screw extruder, 34, 35, 44, 124
Screw press, 44, 45, 114, 115
Solid-state NMR, 26, 27
Sonnergaard Model (Log-Exp-Equation), 67

Spencer and Heckel model, 65
Steam explosion, 136, 137, 141, 144
Sugar Yields, 125–129, 131, 132, 135–137, 157

T
Tablet press, 38, 41, 124
Technoeconomic analysis (TEA), 91, 101, 102
Thermochemical conversion, 4, 9, 15, 24, 44, 123–157
Torrefaction, 56, 136–140, 144, 149, 152, 154

U
Used wood, 2, 171

W
Walker model, 65
Wood processing industry, by-products, and residues, 171
Woody biomass, 2, 6, 10–13, 18, 23, 38, 42, 45, 46, 49, 53, 89, 124, 125, 138, 141, 142, 155, 164, 166, 171–174, 176, 178, 179

The manufacturer's authorised representative in the EU is Springer Nature Customer Service Centre GmbH, Europaplatz 3, 69115 Heidelberg, Germany. If you have any concerns regarding our products, please contact ProductSafety@springernature.com

Printed and bound by CPI Group (UK) Ltd, Croydon, CR0 4YY
28/09/2026
02224092-0013